Human Approaches to the Universe

*Schriften der Luther-Agricola-Gesellschaft 60*

# Human Approaches to the Universe

*INTERDISCIPLINARY STUDIES*

Edited by Eeva Martikainen

Luther-Agricola-Society
2005

Luther-Agricola-Seura
PL 33 (Aleksanterinkatu 7)
FIN-00014 Helsingin Yliopisto
Finland

Distributed by:

Bookstore Tiedekirja
Kirkkokatu 14
FIN-00170 Helsinki
Finland
www.tsv.fi



ISBN 951-9047-94-8
ISSN 1236-9675

Gummerus Kirjapaino Oy
Vaajakoski 2005

# Contents

# Introduction: Human Approaches to the Universe

Eeva Martikainen

It has been the dream of the philosophers of all ages to devise a consistent metaphysic without internal contradiction and in accordance with experience. Reflecting on the huge amount of information provided by his senses, man is confronted with a fundamental question: is this all there is or is there also a reality that cannot be reached by the senses, not even with the aid of sophisticated instruments? Is there a reality which does not obey the physical laws of nature? This is certainly a philosophical question that is answered differently in different metaphysical systems.

The interdisciplinary colloquium[1] on which the articles published in this volume are based was a continuation of the international colloquium held in 2000, also in Helsinki, the theme of which was the history of cosmology and the new challenges it presents.[2] At that time the colloquium dealt in particular with the problem of causality in the Aristotelian tradition of thought preceding the era of modern science, as well as in the changes brought about by physics and specifically by quantum physics. Also, the important question of the relations between physics and metaphysics was then the object of a wider examination based on philosophy

---

[1] Helsinki, Finland 26-27 September 2003.

There are many to whom I am very grateful for their cooperation and help in arranging the colloquium. Tarja Kallio-Tamminen and Inga R. Gammel, members of my research project, have assisted in many ways in the planning and arranging of the colloquium. Lassi Larjo, the secretary of the colloquium, has taken care of numerous practical arrangements. He and Ahti Lampinen have also assisted as redactors of this volume.

[2] See Eeva Martikainen (ed.), *Infinity, Causality and Determinism*. Cosmological Enterprises and their Preconditions. (Contributions to Philosophical Theology, Vol 6.) Peter Lang. Frankfurt am Main 2002.

and its history. Metaphysics has traditionally been the area of philosophy, the first philosophy, that has united knowledge and cogitation. Knowledge is directed towards the real world, while cogitation may also be directed towards merely intellectual objects.

In Antiquity the early natural philosophers provided various models relating to reality. Plato and Aristotle, whose extensive philosophies are often considered to be superior to the pre-Socratic ideas, did not accept the particle-mechanistic conception of nature defended by the atomists. They turned their attention from inanimate nature to living beings and saw that their growth and development could not be understood on the basis of the atomists' idea of "dead" matter. In addition to matter, some kind of organising form was needed.

The modern conception of reality was formed at the dawn of the modern era. It was greatly influenced by classical physics but, as with every wide-ranging theory or framework, it has certain ontological and epistemological presuppositions that necessarily remain metaphysical. Werner Heisenberg claimed that the basic ideas on which modern science was founded come from Antiquity. Galileo united the mathematical approach of Pythagoras with the atomistic and materialistic world-view originating with Democritus and Epicurus.

It was only at the beginning of the modern era that natural philosophy (once again) reached the level of discussion experienced in Antiquity. Classical physics started by abandoning the Aristotelian concept of matter and returned to the idea of dead matter devised by Democritus. As matter was supposed to consist of tiny indestructible particles that obeyed strict deterministic laws, mind, reason and all subjective phenomena had to be stripped from objective reality. Descartes sharply divided reality into two parts: res extensa and res cogitans. He proclaimed that "neither God nor any rational soul present in the world will ever disturb the ordinary course of nature in any way". Physicists began to study the measurable world of extended objects, whereas subjective phenomena were excluded. Descartes' definition of matter was perhaps even more far-reaching than his definition of mind.

Associated with physics there has also been a long tradition of philosophical idealism, the belief that reality is essentially mental in character. The Pythagoreans claimed that mathematical relationships are the underlying reality of nature. The Platonists took nature to be an imperfect reflection of another realm of perfect eternal forms. Both of these themes were expressed in the writings of Kepler and Copernicus at the dawn of modern science. In the eighteenth century Kant and his successors held that the structures of time, space and causality are categories of human thought which we impose on nature; we cannot know things as they are in themselves. The battle between science and humanism has continued unresolved for several centuries.

Since Immanuel Kant the progress in both physics and philosophy has made questions concerning the relation between physics and metaphysics more difficult than previously, but at the same time has made it a challenging subject. Not only quantum mechanics but also the progress in physical cosmology places philosophy and theology in a new situation in which earlier philosophical problems are revived in a new way. All of us are challenged to take part in a multidisciplinary dialogue that is not merely theoretical but most profoundly touches the whole of human life. With the enormous increase in information man's capacity for controlling reality and himself has become more vulnerable than ever. We need the whole of human capacity so that science and the progress of culture might serve their genuine and original purpose: the maintenance and promotion of the life of nature and humankind. The appearance of life is vitally dependent on physical conditions.

Changing the values of various physical constants often results in qualitative changes in complex physical systems, such as the universe. This has sometimes been argued as implying a fine-tuning of the observed physical parameters. In his article "Fine Tuning in the Physical World" *Kari Enqvist* writes that some of these most likely depend on the values of other, more fundamental parameters. Some may attain their values dynamically and are therefore not, even in principle, amenable to even small changes. An example is the total energy density of the universe. He will argue that such uncertainty may well prevent us from attaching any meaningful probability measure to the concept of "small change". The notion of "small change" is, however, relative. It depends on how and over what range the physical parameters can vary. Therefore, before one can make statements about possible fine-tuning, one must be able to answer the question "In what way could things be different?" More properly, one should be able to identify the physical constants which are independent random variables and discover the appropriate probability measure. However, this is not the full story.

What we call a physical parameter is always dependent on our theoretical framework. Nevertheless, we have no obvious way of knowing what are the physical parameters that may vary, nor the way in which they may vary. In fact, instead of asking: "Is the Universe fine-tuned?" one might ask: "If the Universe is not fine-tuned, what would that imply?" Suppose that one day physicists would come and say that the Theory of everything is such and such, and the number of possible universes is such and such a small number. What would it prove, metaphysically speaking? If nothing, then surely the first question also lacks any implications.

*John D. Barrow*, in his article "Our Universe And Others", describes some of the modern motivations for considering our universe in the context of a collection of other possible universes with different properties. This collection can contain universes in which initial conditions, fundamental constants, conserved quantities, and even the dimensions of space and time can be different. He shows how new developments in cosmology have led to the recognition that intrinsically random processes occurring in the early universe may naturally produce a single universe in which these variations occur from one place to another. The implications of this situation for our attempts to understand the universe and the evolution of living complexity within it will be discussed.

*Rainer E. Zimmermann* illustrates the universe in his article "The Modeling of the World as a Glass Bead Game" with Hermann Hesse's novel The Glass Bead Game, which introduces an essentially cosmic game within a symbolic universe which has recently been compared to a "neuronal network of the cosmic mind". In his game all the games known to us today appear to be summarized. Hence, upon closer inspection, the glass bead game shows itself to be a meta-game, a proto-game, and a playful paradigm of playing, at the same time. With its rules it encompasses all that characterizes the reflexive activity of humans within the worldly environment. It thus aims towards all the fields of the sciences and arts which are available as the inventory of human orientation.

Today, the idea of such a playful simulation of the world has become more common than could have been the case in the time of Hesse himself. Our modern attitude to visualizing the world in strict analogy with a computer such as to utilize the concept of the computer itself as a guiding paradigm for envisaging the world, facilitates considerably our understanding of the processes like those which emerge within a playful construction. In this sense, the world appears to us as the programmed result of a matrix which relates to what we call the "world" as a master programme does to a subroutine. And the aspect of simulation does not pose a conceptual problem for us any longer, since we have learned by now that all of our perception is nothing but an interpreting mapping, in the first place, so that both perceptive and cognitive work is very similar to the process of simulation, from the very outset. Hence, it is no longer a question of eventually restricted authenticity. Of course, we have to notice that the computer as a paradigm is not somehow equal to the computers that we are actually capable of constructing. Instead, it is the structural concept of a computer which serves us here as an orientation. The same is true regarding the generalization of the concept of "programme". Because a programme is less something which fixes what can be actualized during its processing, but more the framework of what is possible of being actualized at all. Hence, it is a "field of possibilities" from which the process can actually choose.

*Alfred Driessen*, in his article "Philosophical Consequences of the Gödel Theorem", asks if there is a reality that cannot be reached by the senses, even with the aid of sophisticated instruments? Is there a reality which does not obey the physical laws of nature? This is certainly a philosophical question that is answered differently in different metaphysical systems, but certain experiences support or provide difficulties for certain metaphysical systems. Today the empirical facts have greatly increased and, what is probably more important, these facts are only accessible to a small number of well-prepared specialists. How can a philosopher understand these new empirical facts - often only expressed in the highly formalized language of physicists, mathematicians or other scientists - and evaluate their relevance with regard to his metaphysics? Not to speak about a scientific discussion of the validity of these facts. His attempt is to analyze an important mathematical discovery, the theorem of Gödel, and to demonstrate that the acceptance of this theorem has a direct impact on the consistency of metaphysical systems. It is shown that mathematics is a pointer to a reality that cannot be reached by our senses and that is not subject to physical laws.

To make it very clear, the Gödel theorem is in the field of pure mathematics. The philosopher as a philosopher cannot decide on the correctness of the theorem. What he can do instead is evaluate the general acceptance of this mathematical finding and reflect on the consistency between the consequences of the mathematical theorem and the consequences of his metaphysical view. Three mathematicians are involved in the present argument: first Gödel himself, then the further elaboration by Turing, and finally the consequences for the human mind as worked out by Penrose. As a result, one is encouraged to distinguish two different types of intellectual activity in mathematics, which can both be carried out by humans. The astonishing thing is not the distinction between a formalized, logical approach on the one hand, and intuition, mathematical insight and meaning on the other. What these mathematicians claim, however, is that in principle only one of these approaches can be carried out by devices bound exclusively to the laws of physical reality.

*Wolfgang Hofkirchner, Christian Fuchs* & *Bert Klauninger* have written about the "Informational Universe". From the perspective of up-to-date information science, information systems are no longer only artificially constructed technical systems but rather social or natural systems capable of generating information (and technical information systems are part of them). Information-generating systems, in turn, are self-organising systems. Findings in self-organisation can be generalised on the philosophical level. Generalisations concern the praxeological aspect of world-views: in humans evolution may become a conscious matter; generalisations concern the ontological aspect: evolution is open-ended; and generalisations concern the epistemological aspect: evolution is not fully reconstructable.

*Lassi Larjo*'s theme is the "Intuitive Universe". He asks how we might again have immediacy in human knowledge. Intuition has played a very important role in philosophical systems of Western thought, but its nature and function have always been ultimately inscrutable. Those who speak favourably of intuition seem to need to resort to an inexplicable source of this immediate mode of knowledge. Historically, most influential formulations of the power of intuition can be subsumed under a generic approach called "metaphysics of vision". References to supernatural revelation, divine illumination or to innate forms of knowledge are frequent in this line of argument. A modern scientifically-orientated explanation of the universe finds little systematic use for intuitive knowledge. Although intuition seems to play a modest role in giving birth to innovative new ideas, it does not have any systematic function in human approaches to the universe.

Could this situation be different? Would it be possible to give intuition more credit in our attempts to understand reality? What would a more systematic intuitive approach to the universe be like? In his paper Larjo discusses some major developments in the concept of intuition concentrating on late 18th-century philosophy, especially that of Reid, Kant, and Jacobi, who all in their specific ways used intuition as a means to approach reality.

*Tarja Kallio-Tamminen* develops the ontology of quantum mechanics in her article "Expanding Ontology - Locus for Autonomous, Subjective States in Quantum Framework". She states that the modern conception of reality, whose basic assumptions were formed during the 17th century, has been unable to provide an objective foundation for mental states. Classical physics studied the objective world of material particles obeying strict deterministic laws, and in that context it is difficult to understand how free will or human consciousness might interfere with the predetermined course of the closed material world.

Quantum mechanics has challenged the traditional world-view inherited from Descartes and Newton. Especially the Copenhagen interpretation implied radical reconsideration of classical metaphysics as well as an evaluation of the role of human beings. Niels Bohr abandoned the familiar ideas of determinism, reductionism and the detached observer that were difficult to maintain when explaining quantum phenomena. He considered human beings to be both observers and actors. They are deeply embedded and interconnected with the same reality that their theories aim to describe.

The abstract notion of wave-function refers to a complex quantum-state that is not directly reducible to material particles, even if it gives the probabilities of all possible occurrences in material reality. The new state-description calls for revisions to the established conception of matter and provides an opportunity to reconsider the subtle relation between mind and matter. Mental states may be real and scientifically approachable but the non-local and statistical character of quantum framework does not demand that they be totally identified with brain states. Autonomous, subjective states can be reconciled to the natural world without any need to restrain their unique character.

*Inga R. Gammel* continues the problem of understanding mental phenomena within the framework of quantum mechanics in her article Matter and Myths of the Soul. Perspectives on Plato and Wolfgang Pauli. As she notes, until the seventeenth century matter was still thought to be imbued with soul. However, with the emergence of modern science, matter lost its essence, and soul became a non-question. The physical-psychic problem disappeared from the scientific discussion with the result that the sciences associated themselves with too narrow a concept of reality. Later, at the beginning of the twentieth century, the established scientific world-view came under pressure, due to the discoveries made in micro-physics. Once again it became evident that the age-old questions - such as what is matter, what is soul, and, further, how are they connected? - had not lost their topicality and meaningfulness. Among those who tried to re-establish a scientific discourse on these issues was the physicist Wolfgang Pauli. In addition to his scientific work Pauli made comprehensive studies in philosophy, mythology, religion and psychology. Pauli found inspiration in many sources, past as well as present. Thus Plato's philosophy was important for Pauli, but he also co-operated with Carl Gustav Jung. From this fertile blend emerged a new and broader concept of reality, which included matter as well as World Soul. According to Pauli, reality as such cannot be approached, but we do, however, interact with this unseen reality by means of symbols. Our task, then, is to learn how to translate the symbols in order to materialize their meaning. The phenomenon of synchronicity Pauli regarded as a symbolic event from which man should learn how to grasp its secret meaning.

*Raimo Lehti, Tuomo Suntola* and *Atso A. Eerikäinen* discuss the problem of time in the framework of the theory of relativity. *Raimo Lehti* writes on the einsteinian time in all three «worlds«: in the Microcosmos, Megacosmos and Mesocosmos. We understand by Microcosmos the world of atomic and sub-atomic dimensions, by Megacosmos the world of galactic dimensions, and by Mesocosmos the world 'between'. Einstein developed concepts of time to explain phenomena in all three 'worlds'. Their distinctions may seem problematic as long as we take time itself to be the fundamental concept. But Einstein took as his fundamental the concept of the clock, thus time is defined as 'what the clock shows'. From that perspective, the distinct descriptions of microcosmic, mesocosmic and megacosmic clocks are quite plausible.

*Tuomo Suntola*, in his article, addresses the importance of the difference between the reality which is studied and the models we use to describe it. We are not free to choose the laws of nature but we have considerable freedom in choosing the co-ordinate quantities.Time and distance are the most fundamental co-ordinate quantities. For human perception and logic time and distance are universal for all physical phenomena described. It is a basic rule in all measurements not to change measures for a phenomenon in different environments or circumstances; adjusting the measures is a shortcut, a false way out of the problem. There is no need to sacrifice the constancy and universality of the basic co-ordinate quantities, distance and time, for right predictions of the frequencies of atomic oscillators and the seeming constancy of the velocity of light. In space described as a spherically closed entity with dynamics determined by zero-energy, the balance between motion and gravitation, absolute time and a universal unit of distance can be applied to all phenomena in space.

The choice is as follows: If we fix the velocity of light and make it the maximum speed in space by applying Lorentz transformation as a law of nature, we have to use time and distance as parameters in describing the behaviour of atomic oscillators and measurements of the phase velocity of light in moving frames. We have to postulate the rest energy of matter as a property of mass and accept a relative state of rest as our reference for motion. We are unable to conclude either the overall geometry of space or the development of the geometry with the expansion of space.

If we describe space as a dynamic zero-energy surface of a four-dimensional sphere, time and distance can be used as an absolute co-ordinate quantity and we can understand the dependence of the velocity of light on the development of the universe and on the local gravitational state. We can relate the value of the velocity of light to the total mass and dimensions of space, we can understand why and when the velocity of light is observed as being constant, and also why the velocity of light is the maximum speed achievable in space. Further, we can define the state of rest in any local energy frame and relate that to the absolute state of rest in space. We can identify a quantitative expression to Mach's principle, to the nature of mass, and an energy quantum of radiation.

The curious concept of time in the theory of relativity was originally justified by and based on the relative velocity between the observer and the object. We know for certain that the reading accumulated in the counter of a clock is unambiguously determined by the gravitational and motional state of the clock, and that it is not related to any particular observer. In fact, this was demonstrated in practice for the first time as late as the 1970s, when atomic clocks were taken into aeroplanes and satellites. Until that time all experiments on the frequencies of atomic oscillators were based on observation of the frequency or wavelength of an electromagnetic signal sent by the oscillators. Such experiments combine the inherent frequency of the transmitting oscillator, propagation of the electromagnetic radiation transferring the frequency, reception of the radiation, and the state of a reference oscillator at the observer's location, which can all be handled in absolute time in dynamic space.

*Atso A. Eerikäinen* discusses in his presentation, "Time: Physical and Eternal" the problem of time from the perspectives of metaphysics and physics in order to find a basis for an interdisciplinary definition of time. We may ask whether time is something physical (changing, calculable, measurable) or something eternal (invariable, incalculable, immeasurable). If time is something existing we may equally well ask whether it is something both physical and eternal.These questions are current: they are born in the present situation. What is "now" was an intractable problem for Einstein. Physics and metaphysics are both required to solve this problem.

The Japanese mathematician Hitoshi Kitada, the Canadian biophysicist Peter Beamish and the German theologian Karl Heim have all realized that a holistic understanding of reality presupposes a two-dimensional conception of time. Kitada's Local Time is the physical present moment in the timeless universe created and experienced by the mind. Beamish's Rhythm-Based Time is the cyclic and bi-directional physical present moment, in which the linear "common time" and orthogonal "Kitada Time" unite. Heim's two-dimensional time is the polar and dynamic interaction between physical and eternal time, the energetic information process and its transition. In physical time all change is successive; in eternal time, however, everything occurs at once. Time is in eternity, and eternity is in time.

# Fine Tuning in the Physical Universe

Kari Enqvist

As is well known, a change in the observed values of the physical parameters would often result in a universe unsuitable for man. The consequences have been discussed in detail for example in *The Anthropic Principle* by John Barrow and Frank Tipler.[1] A celebrated example of the anthropic principle concerns the existence of a Carbon state requisite for life on Earth, as first discussed by Fred Hoyle. The issue for Hoyle was: Carbon must be made in stars, but how? It was known that Carbon could be produced in a two step process where first two Helium nuclei collide to form Beryllium. If Beryllium then scatters against Helium, a Carbon nucleus can be created. The problem however is that Beryllium is unstable. In order that Carbon is formed abundantly, the reaction should be very fast and Carbon should be created at every collision. Hoyle realized that this is possible if there exists an energy level of Carbon such that it is just above the combined energies of Helium and Beryllium. In such a case, enhanced (resonant) production of Carbon is possible. We exist, hence a resonant energy level of Carbon must exist, Hoyle concluded. Soon after that the energy level was indeed found in the laboratory.

If short-lived Beryllium did not manage to produce Carbon at each collision with a Helium nucleus, we would not exist. A change in the Carbon energy levels would thus be disastrous for life on Earth. Often it is stated that the values of the physical parameters are indeed fine-tuned to the existence of life, implying that already very small changes would result in a universe totally different from ours, as schematically depicted in Fig. 1.

A typical example of the presumed fine-tuning could be related to the

---

[1] Barrow & Tipler 1986.

value[2] of the neutron mass $m_n$ = 939.57 MeV. The neutron mass is almost equal to the proton mass $m_p$ = 938.27 MeV. Because the neutron is slightly heavier than proton, weak interactions and energetics allows it to decay into an electron (mass $m_e$ = 0.511 MeV) and an essentially massless antineutrino. For historical reasons, the process $n \rightarrow pe\bar{\nu}$ is called beta decay and plays an important role inside the Sun as well as in the early universe. For example, beta decay determines the Hydrogen-to-Helium abundance ratio in the universe. The decay starts when the universe is about one second old and continues until the remaining neutrons are bound in Deuterium. As a consequence, the Helium abundance is 24% and the rest of the baryonic mass is in Hydrogen which can condense to form stars and hence provide for life as we know it.

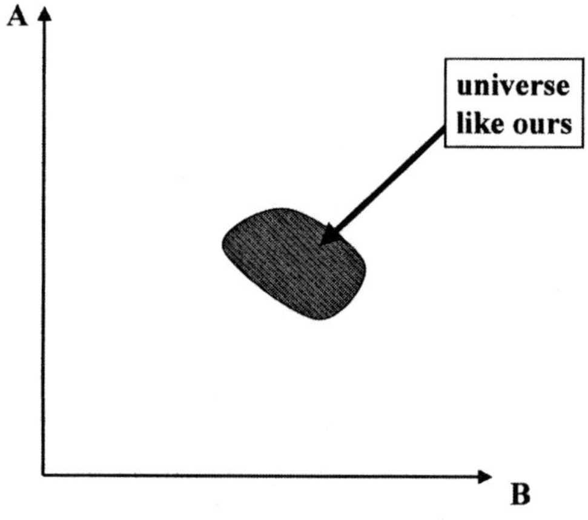


Figure 1. A schematic representation of the possible values spanned by two physical parameters A and B.

[2] Units are such that c=1 throughout this article.

As pointed out by Barrow and Tipler, if the neutron mass were only 0.08% smaller, beta decay would not be energetically possible because the sum of the electron and the proton masses would exceed the neutron mass. In the early universe proton-electron scattering would then turn the protons into neutrons through the process $pe \rightarrow n\bar{\nu}$ and there would be no hydrogen, and hence no stars, and by implication no planets or humans.

A mere change of 0.08% appears strongly to point towards fine-tuning of the neutron mass. But one can easily find even more dramatic examples. By 1980's it was realized that the mass energy density of the universe, $\Omega$, had to take a very special value in the early universe in order to reproduce the observed number. At that time $\Omega$ was poorly known and was thought to lie somewhere in the range $0.1 < \Omega < 2$. However, even such a seemingly wide range of possible variation implies that in earlier times $\Omega$ must have been very close to 1 (Most recent result, due to the WMAP satellite,[3] is $\Omega$=1.02±0.02). This comes about because of the Einstein's equations, which determine the dynamical behaviour of $\Omega$ in the expanding universe. One finds that $\Omega-1 = k/R^2$, where k=0, ±1 is the curvature parameter, and R is the radius of the universe. Plugging in the numbers one finds that at the time when the universe was one second old, $\Omega$ could have differed from 1 only by $10^{-10}$. If one extends the applicability of the Einstein's equations to the maximum, all the way to the Planck time, one would find that then $\Omega$ could have deviated from 1 only less than by a staggeringly small amount of $10^{-60}$.

The issue is not just how to reproduce the quantitative value of $\Omega$ at the present time, but how to get a universe like ours. According to Einstein's equations, if at the Planck time $\Omega$ were bigger than 1 by few times $10^{-60}$, the universe would already have collapsed to a singularity. If at the Planck time $\Omega$ were less than 1 by few times $10^{-60}$, the present universe would be so devoid of matter that there hardly could arise any stars or planets.

Analogous considerations are numerous. For instance, if the temperature fluctuations of the cosmic microwave background, which reflect density perturbations in the mass energy of the universe, were larger than the observed $10^{-5}$ K, one would not have stable planetary orbits; if less, there would be no galaxies.

The notion of "small change" is however relative. It depends on how and in which range the physical parameters *can* vary. Therefore, before one

[3] For details, see WMAP home page map.gsfc.nasa.gov/

can make statements about possible fine-tuning, one must be able to answer the question "In which way things could be different?"

More properly, one should be able to identify those physical constants which are independent random variables and find out what is the appropriate probability measure.

As an example, consider the electron charge, which has the observational value[4] of $e \approx 0.3$. Suppose we now change the value of $e$ to 0.33. Is this a small change?

To answer this question, we should know what is the range within which $e$ can vary. Perhaps we could decide that $e$ can vary between 0 and 1, in which case $0.3 < e < 0.33$ would correspond to a probability of 3%.

However, this is not the full story. What we call a physical parameter is always dependent on our theoretical framework. There is no a priori reason why $e$ should be more fundamental parameter than, say, $e^4$. But if $e^4$ is the randomly varying parameter in the range 0 to 1, finding $0.3 < e < 0.33$ would then correspond to a probability of 0.4%. Who knows, perhaps God is fond of logarithms … or something else.

As a consequence, we cannot reliably determine, say, whether $e \approx 0.3 \rightarrow e \approx 0.33$ would represent a small or large change, or whether any fine-tuning is implied. This is of course a trivial example. Nevertheless, it demonstrates that we have no obvious way of knowing which are the physical parameters that can vary, nor in which way they can vary.

The situation is also complicated by the fact that the physical parameters can be correlated. Their possible range of variation can be constrained dynamically so that all conceivable values are not possible, as illustrated in Fig. 2. For example, the electric charge might not be independent but could, for example, depend on the mass of the proton.

[4] In appropriate units.

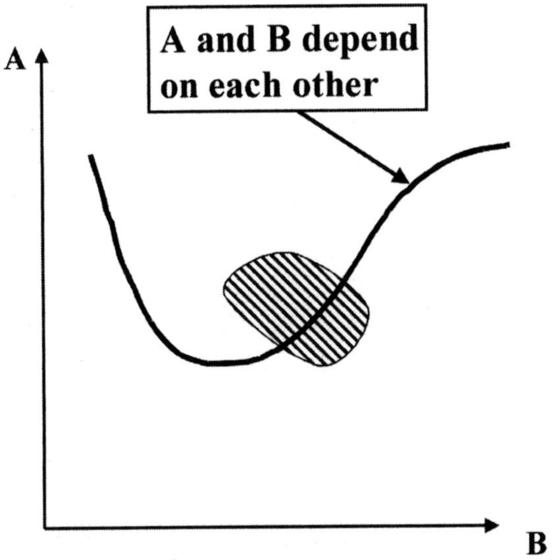


Figure 2. Physical parameters which are correlated. All conceivable values are not possible.

In fact, one of the earliest examples of fine-tuning was resolved by dynamics. Newton and his followers believed that from time to time the Solar System needs divine intervention in order that the planetary orbits can be maintained. It seemed as if Jupiter was slowing down while Saturn was accelerating. For the early Newtonians this was a sign of instability. However, in 1786 Laplace was able to point out that Jupiter and Saturn meet every 59 years.[5] During their close encounters they tug and pull each other, causing orbital changes, and that their roles are reversed every 900 years. Therefore Laplace was able write: "The irregularities of the two planets appeared formerly to be inexplicable by the law of universal gravitation - they now form one of its most striking proofs."

In an analogous fashion, cosmic inflation[6] is a theory which gives a dynamical explanation for Ω being extremely close to 1. According to theory of inflation, the very early universe had a large vacuum energy density -

[5] HAHN 1986.

[6] For a popular account of inflation, see e.g. GUTH 1997.

$H^2$. According to Einstein's equations, a constant vacuum energy causes the universe to expand exponentially with $R \sim \exp(Ht)$. Recalling that $\Omega\text{-}1 = k/R^2$, this implies that $\Omega$ is driven towards 1 exponentially fast. Eventually the vacuum energy decays, and exponential expansion stops. As a consequence, $\Omega = 1$ dynamically and independently of the initial conditions. Note that in order to obtain $\Omega = 1$, no fine tuning of the vacuum energy density is needed.

Inflation thus predicts that, to an extremely high accuracy, $\Omega = 1$. Inflation would not be so popular because of such a simple prediction (or postdiction, depending on one's point of view) alone. What is even more striking is that during inflation quantum fluctuations generate density perturbations with a very well determined spectrum. This is a highly nontrivial prediction. Indeed, WMAP and other cosmic microwave experiments have observed the kind of spectrum of density perturbations predicted by inflation.

The implicit fine tuning in beta decay is also alleviated if one recognizes the fact that the neutron mass is not a fundamental quantity but a consequence of the quark dynamics. The neutron and proton masses are mainly due to the kinetic energies of the quarks and gluons. Only the proton-neutron mass difference can be traced to the different masses of the u and d -type quarks; the proton consists of the three-quark combination uud, the neutron of udd. With $m_u \approx 2$ MeV and $m_d \approx 4$ MeV, $m_n \approx m_p \approx m_d \approx m_u$ (because quarks cannot be observed free, their masses are somewhat uncertain). But this means that a 0.08% change in the neutron mass would imply an almost 50% change in the d-quark mass; changing the kinetic energy of the quarks[7] would modify both the neutron and proton mass values in the same way. Therefore one could well argue that a 0.08% change in the neutron mass, while keeping the proton mass fixed, would represent not a small but a large change.

Neutron is a composite object and naturally composite objects reflect the properties of their constituents. This is a rather trivial way by which physical observables can be constrained. Seemingly unrelated physical parameters could, deep down, be also correlated. This is the situation one finds in unified theories, where the strong and electroweak interactions are unified at some high energy scale. Their relations are determined by the

[7] This would require changing the strong coupling constant.

structure of the theory. As a consequence, the values of the coupling constants are correlated. For instance, the value of the electric charge depends on the value of the strong coupling constant (and hence on the proton and neutron masses). Under certain conditions, such as the requirement that all couplings remain perturbative until the unified scale, one can find that the electric charge we measure cannot be larger than about 0.65.

The situation is much more constrained in theories of everything (TOEs) which unify quantum physics and general relativity. String theories are well known examples. In principle, a TOE contains only three parameters: the Planck mass $M_p$, the Planck constant h, and the velocity of light c. These just set the units, and everything else will be correlated. For instance, the electron charge, neutron mass and the mass of the universe will all depend on each other. Changing *e* would imply, among other things, changing the amplitude of inflationary energy density perturbations.

There are two issues at stake when one discusses the possible changes in the values of the physical parameters. First, what are the consistent sets of laws of nature – i.e. how many TOEs are possible? It seems that not all conceivable laws of nature can be realized; for example, string theories are mathematically consistent only in 1+9 dimensions. It is therefore conceivable that there are only a discrete and finite number of TOEs. Indeed, there could be only one TOE.

The second issue concerns the number of ground states of energy (vacua) within a given TOE. Different ground states would typically correspond to different values of various physical parameters. The laws of nature, as encoded in the structure of the TOE, would also find different realizations in different vacua[8]. For example, some of the symmetries of the theory would be broken at a given vacuum. Possible universes would thus correspond to the (most likely) discrete vacua of the TOE(s), as illustrated in Fig. 3. If the number of possible vacua is small, one could not change the values of the physical parameters arbitrarily and continuously but rather in steps. It is conceivable that there could be only a small number of possible vacua and hence consistent universes[9]. In such a case there would be no small changes whatsoever that one could perform on the physical parameters; instead, there could be only few large changes.

---

[8] These need not be the true ground states; we could be living in a metastable vacuum.

[9] One should however note that currently many theorists believe that the number of string vacua is extremely large.

Be this as it may, the conclusion is that we do not have enough information to claim that the physical, observed universe is fine tuned in some deep, metaphysical sense. A fact is that changes in the physical parameters can give rise to universes that do not resemble ours and which are hostile to life. However, at present we are unable to quantify what is meant by a "small change". We cannot say what is the possible range of physical parameters or to what extent they are correlated. Therefore any utterances on the purpose or meaning of the universe, or on man's special relation to it, based on the perceived fine tunings of some (arbitrarily chosen) physical parameters, are not scientific but rather vehicles for expressing some other sentiments.

In fact, instead of asking: "Is the Universe fine-tuned?" one could ask: "If the Universe is *not* fine-tuned, what would that imply?" Suppose one day physicists would come and say that TOE is such and such and the number of possible universes is such and such small number. What would it prove, metaphysically speaking?

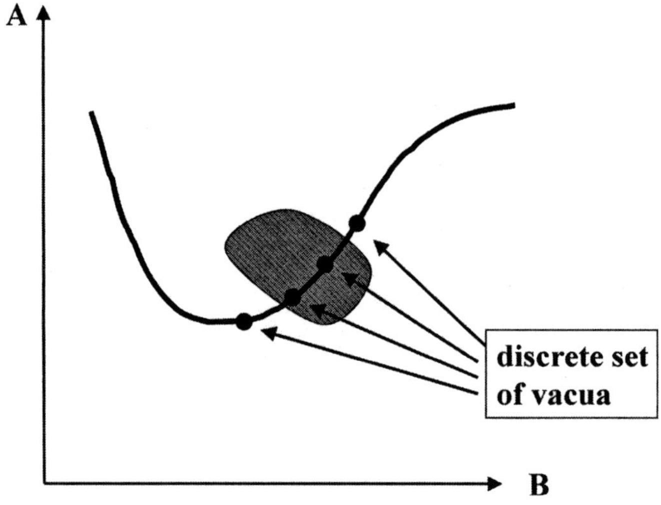


Figure 3. The number of possible worlds (possible values of the physical parameters) could be finite and discrete.

If nothing, then surely the first question is also without any implication.

## *Bibliography*

BARROW, JOHN & TIPLER, FRANK J.
1986 *The Anthropic Principle*. Oxford: Clarendon Press

GUTH, ALAN H.
1997 *The Inflationary Universe*. Reading: Addison-Wesley Publishing.

HAHN, ROGER
1986 "Laplace and Mechanistic Universe" in Lindberg, David C. & Numbers, Ronald L. (eds.): *God and Nature. Historical Essays on the Encounter between Christianity and Science.* Berkeley: University of California Press. 256-276.

# Our Universe And Others

John D. Barrow

## *The Expanding Universe*

The most dramatic consequence of Einstein's general theory of relativity was its prediction that the entire universe should be expanding. By this, we mean that if we could measure the separation of distant clusters of galaxies we would find their separation tomorrow to be greater than it is today. Nevertheless, *we* are not expanding, the Earth is not expanding, nor is our Galaxy. These structures do not expand because they are bound together by stronger local forces of chemical or gravitational origin. Only when we reach the scale of great clusters of galaxies do we find the markers that participate in the expansion of the universe. The first solutions of Einstein's equations which displayed this expansion were found in the early 1920s by the young St Petersburg mathematician and pioneer meteorologist, Alexander Friedmann. These 'Friedmann universes' still provide us with the best description of the overall structure of the universe and its expansion in time.

Seven years after Friedmann's theoretical deductions Edwin Hubble found convincing observational evidence that the universe was indeed expanding. The spectra of light from stars in distant galaxies were shifted systematically in the direction of the red end of the optical spectrum when compared with the same spectral lines measured on Earth. There was a simple interpretation for the systematic shift of lines of all wavelengths towards lower energies. It was analogous to the well-known Doppler shift whereby waves emitted by a receding source are received less frequently than waves emitted by a stationary or approaching source. By measuring the frequency shift it was possible to deduce the recession speed of the light source. By comparing objects of the same intrinsic brightness it was possible to infer their relative distances from us by measuring their apparent brightness. The result was Hubble's famous expansion 'law' showing the speed of expansion of the universe increasing with distance.

The expansion means that the average conditions in the universe are continuously changing. The temperature and density of matter were greater in the past than today. Astronomers have gradually pieced together a pleasingly consistent picture for the universal history that unfolds as the universe expands.

We live about 13.7 billion years after the expansion of the universe appears to have begun unless some unknown physics intervenes in the split second before the beginning is reached to make things different. However, we are able to produce a well-tested model of the early universe back to a time that lies just one second after the apparent beginning. The density of matter is still little more than that of water at that time. Remarkably, when the universe is between one second and three minutes old it behaves like a vast nuclear reactor producing the light isotopes of deuterium, helium-3, helium-4, and lithium-7 in abundances which correspond closely to those observed around the universe today. These detailed astronomical observations allow us to confirm the consistency of our model of the universe right back to these very early times. Likewise, the existence of the microwave background radiation — its thermal spectrum, dipole temperature variation, and flat spectrum of higher-multipole temperature fluctuations (all observed with great precision by the COBE and WMAP satellites and by ground-based and air-borne observations) — confirm the general picture of the hot early universe. Before one second we need a fuller description of the behaviour and identity of elementary particles of matter than we possess at present. When the universe is younger than about $10^{-10}$ seconds, it experiences extremes of temperature and energy that exceed any we can produce on Earth in particle colliders at CERN.

This overall scenario of expansion from a hot early state to a cooler present state is usually referred to as the 'Big Bang' model and is accepted as the working picture for the evolution of the universe by almost all cosmologists. Many aspects of its evolution (even at very late stages) are still uncertain and detailed variants of the Big Bang model are actively explored to discover which can provide a viable explanation for the origin of galaxies.

## *Big and Old, Dark and Cold*

One of the curious features of the universe is the way in which it presents us with an environment which is superficially so hostile to life. But again, appearances can be deceptive. We know that the universe is expanding and therefore its huge size is a consequence of its great age. And any universe that contains the building blocks of complexity must be old enough for stars to form and generate the elements on which such complexity is based. This requires elements heavier than hydrogen and helium which are formed into the first three minutes of the Big Bang. The heavier biochemical elements, like carbon, are made from them by nuclear reactions in the stars. When stars die these biochemical elements are dispersed into space and ultimately find their way into planets and into people. This process of nuclear alchemy is long and slow. It takes billions of year to run its course. Thus a universe that contains 'observers' must be billions of years old and hence billions of light years in size. These are necessary conditions for life to be possible. Further consequences then follow. The large size of a habitable universe ensures that it has a very low average density and so galaxies and stars are widely separated. Outposts of life are likely to be separated by vast astronomical distances, ensuring that development occurs in isolation from other outposts of life, at least until technical knowledge is very sophisticated. The large amount of expansion also ensures that the universe is very cold. This, in turn, means that the night sky appears dark. There is too little energy density in the universe to make it bright. Thus universes that meet the necessary conditions for life are big and old, dark and cold. One might speculate that these aspects of the universes (which should be universal features for observers everywhere) play a significant role in shaping our religious and philosophical impressions of the universe and our place within it. Again, we stress how deceptive appearances can be. Many philosophers have appealed to the vastness and sparseness of the universe as evidence for its basic dysteleological character. Yet the discovery of the expansion of the universe shows how subtle this matter is. Those aspects of the universe which, to some commentators, appear so obviously in conflict with any interpretation of the universe as hospitable for life, turn out to be crucial features that are necessary for a universe to support complexity of any known sort.

It is worth looking a little more closely at the emptiness of space. If we were to smooth out all the matter in the universe into a smooth sea of evenly-spaced particles, there would be only about one atom in every cubic metre of space — a far better vacuum than could ever be created in a laboratory on Earth. If we gather the matter into clumps then this density corresponds to about one Earth-sized planet in every cube of side 10 light years, one star in every cube of side 1000 light years, one galaxy of 100 billion stars in every cube of side 10 million light years. Faced with these numbers we begin to see why there seem to be so few extraterrestrials about: in a universe that is big enough and old enough to give rise to life, the average distances between planets and stars are necessarily huge. In any expanding universe there is a simple approximate relationship $Grt^2 \approx 1$ between the age of the universe, t, and the density of matter, r, mediated only by Newton's constant of gravitation, G. This relation[1] ensures that there is an unbreakable link between local environmental conditions in the universe and its global structure.

## *Simple Inflation*

The expansion of the universe is delicately poised, very close to the critical dividing line that separates universes which are expanding fast enough to keep going forever from those which will ultimately contract back towards a cataclysmic Big Crunch in the future. Indeed, so close are we to this critical divide that our observations cannot tell us for sure what the long-range forecast holds. However, it is the very proximity of the expansion to the watershed that is the big mystery: a priori it seems highly unlikely to arise by chance. Again, this is not totally unexpected: universes that expand too fast are unable to aggregate material into galaxies and stars, so the building blocks of complex life cannot be made. By contrast, universes that expand too slowly end up collapsing into contraction before the billions of years needed for stars to form have passed. Only universes close to the critical divide can live long enough and expand gently enough for the stars and planets to form

Since 1980, there has been an explanation for the universe's proximity to the critical divide and its very large size.[2] These are features that can be

[1] Whitrow 1956.

explained by a sequence of events that may be very probable in any type of universe, no matter how it starts out expanding. This so called 'inflationary' theory of the very early universe introduces a slight gloss on the simple picture of an expanding universe. The standard Big-Bang picture of the expanding universe, that has been with us since the 1920s, has a particular property: the expansion is decelerating. No matter whether the universe is destined to expand forever, or to collapse back to a Big Crunch, the expansion is always decelerated by the gravitational attraction of its material contents.

It had always been assumed that gravity is always attractive. But in the 1970s, particle physicists began to find that their theories predicted the existence at high energies of many new forms of matter, called *scalar fields,* whose gravitational effect upon each other was repulsive.[3] If those fields were to become the largest contributors to the density of the universe at some stage in its very early history, then the deceleration of the universe would be replaced by a surge of acceleration. Remarkably, it seems that if these scalar fields do exist, then they invariably come to be the most influential constituent of the universe, and their influence only ceases when they decay away into ordinary matter and radiation.

The inflationary universe hypothesis is simply that such a brief period of *accelerated* expansion occurred early in the history of the universe. This brief inflationary episode sounds innocuous but it can solve many long-standing cosmological problems. A past period of accelerated expansion enables us to understand why our visible universe is expanding so close to the critical divide that separates open and closed universes. The fact that we are still so close to this divide, after about fifteen billion years worth of expansion, is quite fantastic. Since any deviation from lying precisely on the critical divide grows steadily with the passage of time, the expansion must have begun extraordinary close to the divide in order to remain so close still today — so close that we still do not know on which side of the divide we lie. But the tendency of the expansion to veer away from the critical divide is just another consequence of the attractiveness of the gravitational force. But if gravity is repulsive and the expansion accelerates, it will drive the expansion ever closer to the critical divide. If inflation lasted for long

[2] GUTH 1981, 347.

[3] BARROW 1994.

enough, from $10^{-35}$ sec to$10^{-33}$ sec is enough, then it can explain why our visible universe is still so close to the critical divide.

Another by-product of a short bout of cosmic acceleration is that any large irregularities in the density of the universe get ironed out and the expansion very quickly goes at the same rate in every direction, just as we see today. This explains a property of the expansion of the universe that has always struck cosmologists as mysterious and unlikely — because there are so many more ways for the universe to expand at different rates in different directions. If inflation occurred, the whole of our visible universe can arise from the expansion of a region that is small enough for light signals to cross at very early times. Any irregularities get ironed out very quickly. In the old, non-inflationary Big Bang theory the situation was very different. Our visible part of the universe had to emerge from a region vastly bigger than any that light rays could smooth out. It was therefore a complete mystery why our visible universe looks so similar in every direction on the sky to within one part in 100,000 — as observed.

In reality, the tiny region which grew into our visible universe could not have started out perfectly smooth. That is impossible. There must always be some tiny level of quantum statistical fluctuation present. Remarkably, a period of inflation stretches these fluctuations to very large astronomical scales, where they appear to have been seen by the COBE and WMAP satellites. So far the high-precision measurements made by WMAP are in remarkable agreement with the detailed predictions of the simplest and most natural inflationary universe scenario. Future satellite data from WMAP and the European Planck Surveyor Mission will reduce the experimental yet further. It is remarkable that these observations are providing us with a direct experimental probe of events that occurred when the universe was perhaps only about $10^{-35}$ seconds old.

## *Eternal and Chaotic Inflation*

Beyond the boundary of the little patch of the early universe that inflated to encompass the whole of our visible universe lie many (perhaps infinitely many) other such causally linked patches that can all undergo varying amounts of inflation to produce extended regions of our universe that lie beyond our visible horizon today. This leads us to expect that our universe possesses a complex spatial structure and the conditions that we can see within our visible horizon, about fifteen billion light years away, are unlikely to be typical of those far beyond it. This complicated picture is usually termed 'chaotic inflation'.[4]

It has always been appreciated that the Universe might have a different structure beyond our visible horizon. However, prior to the investigation of inflationary universe models this was always regarded as an overly positivistic possibility, often mooted by pessimistic philosophers, but which had no positive evidence in its favour. The situation has changed: the chaotic inflationary universe model gives a real reason to expect that the geography of the Universe beyond our horizon differs from that in the visible part of the Universe.

It was then realised by Alex Vilenkin and Andre Linde that the situation is probably even more complicated. If a region inflates then it generally creates within itself, on minute length scales, the conditions for further inflation to occur from many of its parts. This process can continue into the infinite future, with inflated regions producing further sub-regions which inflate, and so on... *ad infinitum*. The process has no end. It has been called the 'eternal' or 'self-reproducing' inflationary universe.[5] As yet, it is not known whether it need have a beginning.

How likely is it that a large life-supporting piece of the Universe like the region we can see arises in this eternal process? So far, nobody knows. The problem of defining unambiguously what is meant by probability in this process remains an unsolved problem.

## *The Origami of the Universe*

The phenomenon of inflation is expected to occur when the universe is very young, just about $10^{-35}$ second old. Such small times seem bizarre. Clearly, anthropomorphic units like 'seconds' were not designed for cosmological problems. There are other humanly devised measures of time which might help alleviate the failure of our imaginations but they would still fail to give a feeling that we had got the right superhuman perspective

---

[4] BARROW & TIPLER 1986.

[5] LINDE 1994.

on cosmic time. Fortunately, there is a measure of time that is intrinsically defined by the forces of Nature that enables us to create a better perspective. In 1870, an Irish physicist, George Johnstone Stoney (also famous for both naming and predicting the value of the charge on an electron[6]) realised that three of the constants of Nature, e, G, and c, could be combined to create units of mass, length, and time that were independent of human standards. Thirty years later, Max Planck rediscovered this idea and created 'natural units' using the constants of Nature[7] h,c, and G. The resulting Planck-Stoney units of length, $L_{pl}$ ,and time, $T_{pl}$ , are the only quantities with units of length and time that can be made from these fundamental constants which govern quantum, relativistic, and gravitational phenomena. They are:

$$L_{pl} = (Gh/c^3)^{1/2} = 4.1 \times 10^{-33} \text{ cm}$$

$$T_{pl} = (Gh/c^5)^{1/2} = 1.3 \times 10^{-43} \text{ sec}$$

Theses scales of distance and time mark the boundaries below which our current theories of gravity and quantum reality fail. They must be merged to create a new theory of quantum gravity. In effect, at times earlier than $10^{-43}$ sec the entire universe is dominated by quantum mechanical uncertainty. No one knows what the structure of space will be like on scales smaller than $L_{pl}$, perhaps knotted or interconnected in some complicated way. However, it is fortunate that this problem arises well before we expect inflation to occur.

In these natural units the visible universe is now $10^{60}$ Planck times old and $10^{60}$ times the Planck length is size. This is an objective way of saying that the universe is big and old without recourse to comparisons with human measuring artefacts. One way of looking at what inflation has done for the universe is to regard the Planck length as the natural size for a space under the influence of gravitational forces. Inflation can make the universe become much bigger, in our case $10^{60}$ times bigger.

These numbers are still very large and it may be helpful to bring them closer to the imaginable in the following way. Suppose we take a piece of A4 paper and fold it in half, then in half again. You may be surprised to find that you can't fold it more than seven times. Halving moves very

---

6 Barrow 1983a, 24.

7 Barrow 2002.

quickly. If you could have folded the paper in half 30 times it would already be the size of a single atom; fold it 47 times and you are smaller than an atomic nucleus; fold it 114 times (a lot but not unimaginable) and you are down to the Planck length where quantum gravity (and quantum origami) rules. To make the link with the scale of the whole visible universe imagine doubling the size of the A4 paper 90 times and it is about $10^{27}$ centimetres in size. Just 204 paper-foldings divides the smallest and the largest dimensions of space in the physical universe.

## *What happens at the Planck scale?*

The Planck scale is the final frontier for cosmologists at present. If we try to reconstruct the state of the universe earlier than the Planck time then we encounter an environment where no known laws apply for sure. The entire universe is beset by quantum uncertainty. Of course, this does not stop anyone from trying and there are many speculative scenarios which attempt to unravel the complexities of the Planck epoch and discover some residue of that era that might be observable today. We find mathematical studies of how the universe might have emerged quantum mechanically from some suitably defined state or greater simplicity, or even of 'nothing'. One of the stimuli for such investigations is the feature that the energy of the universe is zero (Einstein's equations ensure that the negative potential energies of attraction between all masses exactly cancel out the positive contributions from their masses and motions). New cosmologies have been proposed in which the expanding universe is a consequence of colliding sheets ('branes') of energy. These ideas entail some aspects of the older models of 'bouncing' universes which undergo a sequence of expanding and contracting phases' like a bouncing ball. But yet more radical possibilities should be taken seriously. The familiar distinction between space and time should not be assumed to survive the Planck epoch, nor indeed should the simple picture of a fixed number of dimensions to space and time. These dimensions may change, or even cease to have a clear meaning at the Planck time. The most intensively studied superstring theories, and their successor M-theory, predict that there are more dimensions of space than the three with which we are familiar.

## *Extra dimensions*

During his early career, the great German philosopher, Immanuel Kant, was more interested in science than in philosophy. He was a great admirer of Newton and his laws of gravity and motion and applied himself to understand them in different ways and to apply them to astronomical problems like the origin of the solar system. Kant's most imaginative idea, sparked by Newton's work, was to pose the question 'why does space have three dimensions?'.[8] Kant had noticed a very profound thing: that Newton's famous inverse-square law of gravity was intimately connected with the fact that space has three dimensions. If space had four dimensions then there would be an inverse-cube law, if it had 100 dimensions there would be an inverse $99^{th}$ power law of gravity. In general, an N-dimensional world exhibits a force law for gravity and electromagnetism which falls off as the $(N\text{-}1)^{st}$ power of distance. This is why we see inverse-square force laws in Nature.

The nature of universes with different dimensions of both space and time has been explored by a number of scientists.[9] We can assume that the laws of Nature keep the same forms but permit the numbers of dimensions of space and time to range freely over all possibilities.

The chequer-board of all possibilities can be whittled down dramatically by the imposition of a small number of reasonable requirements that seem likely to be necessary for information processing and 'life' to exist. If we want the future to be determined by the present then we eliminate all those regions of the board marked 'unpredictable'. If we want stable atoms to exist along with stable orbits of bodies (planets) around stars then we cut out the strips marked 'unstable'. Eliminating worlds in which there is only faster-than-light signalling we are left with our own world of 3 + 1 dimensions of space plus time along with very simple worlds that have 2 + 1, 1 + 1, and 1 + 2 dimensions of space plus time. Such worlds have no gravitational fields and are usually thought to be too simple to contain living things.

Worlds with more than one time are hard to imagine and appear to offer many more possibilities. Alas, they seem to offer so many possibilities for happenings that the elementary particles of matter are far less stable

---

[8] HANDYSIDE 1929.

[9] BARROW 1983b, 337; BARROW & TIPLER 1986, chap. 4.

than in worlds with a single time dimension. Protons can decay easily into neutrons, positrons and neutrinos and electrons can decay into neutrons, antiprotons and neutrinos. The overall effect of extra time dimensions is to make complex structures highly unstable unless they can exist at extremely low temperatures[10]. Worlds with more than one time dimension do not allow the future to be predicted from the present. In this sense they are rather like worlds with no time dimension. A complex organised system, like that needed for life, would not be able to use the information gleaned from its environment to inform its future behaviour. It would remain simple, too simple to evolve.

The direct anthropic link between the number of dimensions of space and the existence of living observers was first made by the English cosmologist Gerald Whitrow, in 1955. Asking the question *'Why do we observe the universe to possess three dimensions?'*, he sought to provide a new type of answer[11],[12] by arguing that thinking observers could only exist in three-dimensional worlds. Indeed, he suggested that it would be possible to deduce the dimensionality of world from the fact that we, or another form of intelligent life, exist. Whitrow's approach is the first modern application of what would now be called an 'Anthropic Principle'.

String theories are the only current theories of physics which do not lead to internal contradictions or to predictions that measurable quantities have infinite values when gravity is merged with the other forces of Nature. Yet these appear to require the universe to have many more dimensions of space than the three that we habitually experience. Since we see only three dimensions we must either conclude that these theories are wrong, that dimensions can be something other than what we are used to thinking them to be, or that lots of dimensions of space are hiding somewhere. While either of the first two options might turn out to be the case, it is generally assumed that the third provides the answer to the conundrum. Some process must be found which allows three (and only three) of the total number of dimensions of space to grow very large while the rest remain trapped at the Planck scale of size, where their effects are imperceptible to us. The conundrum is how three of them have become so much bigger — $10^{60}$ times bigger — than the Planck size. What is required is a process which leads to

[10] DORLING 1969, 539; YNDURAIN 1991, 15.
[11] WHITROW 1955, 13.
[12] WHITROW 1959.

the inflation of only three of the dimensions. At present no such selective process is known. It might be random in character, so that its choice of three large dimensions was not programmed into the laws of physics. Alternatively, there might be a deep reason why three and only three dimensions can inflate.

Putting the mystery of the selective inflation process to one side, we see that we are confronted with a major uncertainty. The true constants of Nature, and the forms of the laws of Nature, are really framed in more dimensions than three. The quantities that we call the constants of physics are just three-dimensional shadows thrown by the true constants. Remarkably, if the extra dimensions exist and change their size by expanding as our three-dimensional part of the universe does, then this would be revealed by a change in our 'constants' of Nature at exactly the same rate.[13]

One other escape from the consequences of extra dimensions and variations in constants that are seen to be constant to very high precision is currently being explored in great detail. Suppose only the gravitational force 'sees' and influences *all* the dimensions of space. The other forces — electromagnetism, radioactivity, and nuclear — only act in three of the dimensions (the 'braneworld') because they derive from strings whose endpoints are fixed on the brane surface. This 'braneworld' picture shows why gravity is so much weaker than the other forces of Nature.

## *Changing Constants*

The realisation that the constants of nature we observe in three dimensional space may not be fundamental, or constant, has renewed interest in testing their constancy by high-precision observations. During the last two years John Webb, Michael Murphy, Victor Flambaum, Vladimir Dzuba, Chris Churchill, Jason Prochaska, Art Wolfe, and I have applied a new technique to analyse absorption spectra from distant quasars. We look at the separation between lines caused by the absorption of quasar light by different chemical elements in clouds of dust in between the quasar and us. These separations depend sensitively on the value of a particular constant of Nature, the fine structure constant, $\alpha$, at the redshift where the absorption occurs. By comparing with their separations here and now in the lab-

[13] Marciano 1984, 489; Barrow 1987, 1805.

oratory we have a probe of whether $\alpha$ can have over twelve billion years. This method is more than 500 times more sensitive than any other experimental probe of $\alpha$ and exploits the fact that a change in a affects separations between relativistic spectral lines in different species in very distinctive ways. A small decrease in $\alpha$ in the past will cause some separations between lines to increase and others to decrease. By using computational solutions of the equations of atomic structure we can determine the shifts in line spacings that would result from tiny changes in $\alpha$ and find the shift, $\Delta\alpha$, that best fits all the observed line separations.

This method, which we called the many multiplet (or MM) method is far more sensitive than the other astronomical methods. It has now been applied to observations of 147 quasars, looking at separations between their magnesium, iron, nickel, chromium, zinc, and aluminium lines.

The results gathered and analyzed over two years have proved to be unexpected and potentially far reaching. We find a persistent and significant difference in the separation of spectral lines at high redshift compared to the separation of the same lines when measured in the laboratory. The complicated 'fingerprint' of shifts matches that expected if the value of the fine structure constant was *smaller* by about seven parts in a million at the time when the absorption lines were formed. If we take the observations of sources lying between redshifts of z = 0.5 -3.5 as a whole, the observed shift is[14]

$$\Delta\alpha/\alpha \equiv [\alpha(z) - \alpha(\text{now})]/\ \alpha(\text{now}) = -0.72 \pm 0.18 \times 10^{-5}.$$

A change in the fine structure constant of this magnitude might have had adverse consequences 1.8 billion years ago (equivalent to a redshift of about 0.1) on Earth. We have geochemical data to show that a natural nuclear 'reactor' ran intermittently below the Earth's surface in Oklo, a region of Gabon in West Africa[8]. These natural reactions only occurred because of a series of remarkable geological and nuclear coincidences. A particular nuclear energy level which determines the probability of neutron capture by samarium must have remained the same to high precision between the operation of the natural reactor and the present. This energy level depends on the value of the constant $\alpha$. The Oklo events imply $\Delta\alpha/\alpha < 10^{-7}$ when

[14] WEBB & AL. 2001; MURPHY & AL. 2001, 1208.

the reactor ran.[15] However, whereas Oklo limits changes in a about 2 billion years ago, the quasar observations span the range from about 3 to 11 billion years ago. Simple theories[16] which include varying a are consistent with both limits because the variations cease when the universal expansion starts to accelerate at $z \approx 0.5$-$0.7$. Before that it is predicted to increase slowly, as the logarithm of time, back to when the universe was about 300,000 years old, and to be constant before that.

More important we find that if we fit the one free defining parameter in the varying-$\alpha$ theory so as to fit the quasar data as well as possible, it predicts that we should observe a violation of Einstein's weak equivalence principle (that masses of different composition should fall with the same acceleration under gravity in a vacuum) at a level of one part in $10^{13}$. Existing observations place an upper limit of a part in $10^{12}$ on any such violation, so a small increase in sensitivity, well within the reach of future space-based observations, could offer a critical independent test of the astronomical evidence for varying $\alpha$.

## *Braking the Changes*

If the constants of Nature are slowly changing then life could be on a one-way slide to extinction. We know that our existence hinges upon many peculiar coincidences between the values of different constants of Nature and the observed values of the constants fall within some very narrow windows of opportunity for the existence of atomic complexity. If the values of these constants are actually shifting, could they slip out of the range that allows life to exist?

There are three overall trajectories for an expanding universe to follow. The 'closed' universe expands too slowly to overcome the decelerating effects of gravity and eventually it collapses back to high density. The 'open' universe has lots more expansion energy than gravitational deceleration and the expansion runs away forever. The in-between world, that is often called the 'flat' or 'critical' universe, has a perfect balance between expansion energy and gravity and keeps on expanding for ever. Remarkably, our universe is tantalisingly close to this critical or 'flat' state today. This is a

[15] FUJII & AL. 2000, 38.

[16] SANDVIK, BARROW & MAGUEIJO 2002.

feature of the universe that inflation neatly explains as one of the consequences of a bout of accelerated expansion in its very early stages. However, recent observations have shown that our universe expansion may have begun to accelerate quite recently, when the universe was about three-quarters of its current size. This is expected if the universe contains a constant 'vacuum' energy density described by the famous cosmological constant of Einstein. But why it should have a value that allows it to dominate the expansion for the first time just a few billion years ago remains a mystery.

The overall pattern of the expansion is telling us that about 70% of the energy in the universe is in the vacuum form that acts to accelerate the expansion whilst almost all the rest is in the form of matter, only about 2% of which is luminous.

What happens to this story if the fine structure constant changes? We have found that the fine structure constant only changes when there are no deviations from the critical expansion and the vacuum energy is absent. During the radiation era there is no significant change at all. But once the matter era begins, when the universe is about 300,000 years old the fine structure constant starts to *increase* in value very slowly, logarithmically in time.[17] When the curvature era begins, or if the vacuum energy begins to accelerate the universe, the increase stops: a remains constant.

We know that if the fine structure becomes too large then atoms and molecules will be unable to exist and no stars will be able to form because their centres will be too cool to initiate self-sustaining nuclear reactions. It is therefore crucial that the dust era of cosmic history during which the fine structure constant increases does not last too long. Without the vacuum energy or the curvature to stop the steady increase in the fine structure constant's value there would come a time when no life is possible. The universe would cease to be habitable by atom-based forms of life who relied upon stars for energy

It has always been something of a mystery why our universe is so close to the critical state of expansion today and why the vacuum energy is so fantastically small. We know that if we were too far from the critical expansion then life would have been far less likely to have evolved on Earth, and would probably be impossible anywhere else in the universe as well. If universes are too curvature dominated then the expansion goes so fast that islands of material cannot overcome the effect of the expansion and contract

[17] BARROW, SANDVIK & MAGUEIJO 2002a.

to form galaxies and stars. On the other had, as we have seen, if the universe expands too slowly it soon collapses back to a Big Crunch. Dense islands of material form too quickly and fall into large black holes before stars and biochemistry have a chance to form. Likewise, with the vacuum energy. If it were ten times bigger it would have started accelerating the universe so early on in its history that galaxies and stars would not have been able to separate out from the overall expansion and we would not be here if they were. But the possibility of varying constants provides us with a possible reason why the universe could not be observed by us to be exactly critical and to have no vacuum energy.[18] The vacuum energy and the curvature are the brake-pads of the universe that turn off variations in the constants of Nature. They stop the constants changing. If they are not stopped then a will reach values that prevent the existence of atoms, nuclei, planets and stars. The universe will eventually become lifeless, unable to contain the building blocks of complexity. Life, like all good things, would have to come to an end.

## *Bibliography*

BARROW, JOHN D.

1983a "Natural Units Before Planck" in *QJRAS*, 24.

1983b "Dimensionality" in *Philosophical Tranactions of the Royal Society* A 310, 337.

1987 "Observational Limits of the Time-Evolution of Extra Spatial Dimensions" in *Physical Review D*, 35, 1805.

1994 *The Origin of the Universe*. London: Orion.

2002 *The Constants of Nature: from Alpha to Omega*. London: Jonathan Cape.

BARROW, JOHN. D., SANDVIK, H. & MAGUEIJO, J.

2002a "Behavior of varying-alpha cosmologies" in *Physical Review D*, 65, 063504.

2002b "Anthropic reason for nonzero flatness and λ" in *Physical Review D*, 65, 123501.

BARROW, JOHN D. & TIPLER F. J.

1986 *The Anthropic Cosmological Principle*. Oxford: Oxford University Press.

DORLING, J.

1969 "The Dimensionality of Time" in *American Journal of Physics*, 38, 539.

[18] BARROW & SANDVIK & MAGUEIJO 2002b

FUJII, Y & AL
2000 "The Nuclear Reactional Okloz Billion Years Ago" in *Nuclear Physics* B 573.

GUTH, A.
1981 "Inflationary Universe: A possible solution to the horizon and flatness problems" in *Phyical Review D*, 23, 347.

KANT, IMMANUEL
1929 *Kant's Inaugural Dissertation and Early Writings on Space.* Transl. by Handyside, John. Chicago: University of Chicago Press.

LINDE, A
1994 "The Self-Reproducing Inflationary Universe" in *Scientific American*, no. 5, vol. 32.

MARCIANO, W.
1984 "Time variation of the fundamental 'constants' and Kaluza-Klein theories" in *Physical Review Letters*, 52, 489.

MURPHY, M. & AL.
2001 "Possible evidence for a variable fine structure constant from QSO absorption lines: motivations, analysis and results" in *MNRAS*, 327, 1208.

SANDVIK, H. & BARROW, JOHN D. & MAGUEIJO J.
2002 "A Simple Cosmology with a Varying Fine Structure Constant" in *Physical Review Letters*, 88, 031302.

WEBB, J. K. & AL.
2001 "Further evidence for cosmological evolution of the fine structure constant" in *Physical Review Letters*, 87, 091301.

WHITROW, G.
1955 "Why space has three dimensions" in *British Journal for the Philosophy of Science*, 6, 13.
1956 "Private Communication to Mascall, recorded " in Mascall, E. L.: *Christian Theology and Natural Science*. London: Longmans.
1959 *The Structure and Evolution of the Universe*. London: Hutchinson.

YNDURAIN, F.
1991 "Disappearance of Matter due to Causality and Probability Violations in Theories with Extra Timelike Dimensions." in *Physics Letters B* 256.

# The Modeling of Nature as a Glass Bead Game

RAINER E. ZIMMERMANN

## *I*

In his novel *The Glass Bead Game*, Hermann Hesse introduces an essentially cosmic game within a symbolical Universe which recently has been compared with a „neuronal network of the cosmic mind".[1] In his game all those games which are known to us today appear to be summarized. Hence, by closer inspection, the glass bead game shows up as a *meta-game*, a *proto-game*, and as a *playful paradigm of playing*, at the same time. As its rules it encompasses all what characterizes the reflexive activity of humans within the worldly environment. It thus aims towards all those fields of the sciences and arts which are available as the inventory of human orientation. Hesse describes the rules as such: „These rules, the sign language and grammar of the game, represent a kind of highly developed secret language, in which several sciences and arts, in particular mathematics and music (or the science of music) participate which are able to express the contents and results of practically all of the sciences and relate them to each other."[2]

This systematic form of approach to a universal game, conceptualized in a global manner, implies an important methodological consequence. Hesse continues therefore: „The glass bead game is thus a game with all the contents and the values of our culture, it plays with them, as in the heyday of the arts a painter may have played with the colours on his palette. What humankind has produced in its creative epochs in terms of knowledge, noble thoughts, and works of art, what the successive epochs of learnt reflex-

---

[1] Cf. www.glassbeadgame.com

[2] HERMANN HESSE: *Das Glasperlenspiel* (1943). Quoted here according to the two volume jubilee edition for Hesse's hundredth birthday (1977, I, 12 sq.). Here and later always my translation.

ion have conceptualized and claimed as intellectual property, all of this extraordinary material of intellectual values is being played by the glass bead player like an organ is played by the organist, and this organ is of a hardly graspable perfection, its manuals and pedals are scanning the whole spiritual cosmos, its registers are almost uncountable, theoretically, the complete intellectual contents of the world could be reproduced by playing."

Today, the idea of such a playful simulation of the world has become more common than this could have been the case in the time of Hesse himself. Our modern attitude to visualize the world in strict analogy with a computer such as to utilize the *concept of the computer* itself as a guiding paradigm of grasping the world, facilitates it considerably to understand processes as those which emerge within a playful construction. In this sense, the world appears to us as the programmed result of some matrix which relates to what we call „world" as a master programme does to a subroutine. And the aspect of simulation is not a conceptual problem for us any more, since we have learned by now that *all of our perception* is nothing but an interpreting mapping, in the first place, so that the perceptive as well as cognitive work is very similar to the process of simulation, from the outset. Hence, there is no question of an eventually restricted authenticity anymore.[3] Of course, we have to notice that the computer as a paradigm is not somehow equal to those computers we are actually capable to construct and built. Instead, it is the *structural concept of a computer* which serves us here as an orientation.[4] The same is true as to the generalization of the concept of „programme". Because a programme is less something which fixes what can be actualized during its processing, but more the *framework of what is possible of being actualized* at all. Hence, it is a „field of possibilities" from which the process can actually choose. With a view to the organ metaphor Hesse can continue therefore: „These manuals, pedals, and registers [of the organ] have been fixed by now, any changes or attempts to perfect their number or order are possible only in theory: The enrichment of the game language by means of introducing new contents is delegated to the

[3] Hence, there cannot be something like an ontological proof for the existence of human consciousness. This is so despite the elegant construction of proving one's own authentic existence as given in Stanislaw Lem's famous novel „Solaris": See e.g. the German edition of 1983, dtv: München, 57 sqq. with the description of Lem's *experimentum crucis*. As far as I can see the conclusion there is not completely flawless.

[4] Recent movies like „The Thirteenth Floor" or „Matrix" give a clear impression of this.

possibly strictest control by the superior game's steering committee. But within this fixed construction or, for staying in our picture, within the complex mechanics of this giant organ, the single player is given a whole world of possibilities and combinations, and that among thousand games performed according to the rules only two of them are more similar to each other than merely at their surface, is practically outside of the range of possibility. Even if it happened once by coincidence that two players would choose the same selection of topics as contents of the game, these two games could look and run completely differently according to the style of thinking, the character, mood, and virtuosity of the player."[5]

As Hesse describes in his novel the glass bead game as result of a long development, he sheds some light onto the sciences and arts as they are available to us today exhibiting a permanent striving for epistemic unity: „ ... each attempt to approach the unity of the exact and of the more liberal sciences, each attempt to reconcile science and art or science and religion is based on the same eternal idea which has gained its shape for us in the glass bead game."[6]

Hesse does not even flinch from introducing those fields of human thinking into the game which cannot be grasped and expressed in clear communicative terms of some linguistic discourse, as it is otherwise the case for the sciences and most of the arts. He states: „As the glass bead game had been grown from its beginning in terms of technique and scope up to the infinite, and had become a high art and science in itself, as far as the intellectual claim of the player is being concerned, there was nevertheless something missing. Up to then, in fact, each game had been a chaining, ordering, grouping and confronting of concentrated ideas from many fields of thinking and of beauty, a quick remembering of eternal values and forms, a short masterly flight through the empires of the mind. It was not before considerably later that from the intellectual inventory of the educational system, and in particular from the habits and customs of the Riders of the East, the concept of contemplation came into play ... This was the turn against the religious. It was not anymore the point to simply follow the series of ideas and the complete spiritual mosaic of a game with alert concentration and practised memory in an intellectual manner, but now the demand emerged for a deeper and emotional devotion. After each sign

[5] Hesse 1977, 12.
[6] Hesse 1977, 13.

now which the respective director of the game had conjured up, this sign was being made the object of a strict and silent consideration in terms of its contents, origin, and sense, which forced every player to make the sign's aspects themselves intensively and organically present. The technique and practise of contemplation was exercised by all members of the order and the playing guilds from the élite schools where the art of contemplating and meditating was attended to with greatest care. This was the reason for saving the hieroglyphs of the game from degenerating into mere letters."[7]

Hence, the game shows up as a process which unfolds under two chief aspects, the one aiming at the universal which is part of the social collective, the other aiming at the singular which is represented by the individual. The first aspect refers to the constellations of thinking which are transported by cultural tradition: „In the designations, keys, signatures, and abbreviations of the game language a formula of astronomical mathematics, the form principle of an ancient sonata, an aphorism of Kungfutse and so forth had been stored. A reader who should not be acquainted with the glass bead game can visualize such a scheme as being similar to the scheme of a game of chess, only that the meanings of the figures (pieces) and the possibilities of their relations among each other, and of their interactions, would have to be thought of as being multiplied, and each figure (piece), constellation, move would be associated to an actual contents symbolically designed by the chosen configuration."[8] The second aspect aims at the practical introspection of the person playing: „The formal game aimed at a harmony and an as dense as possible, unbroken, formally perfect unity to be composed by the material contents of each game, the mathematical, linguistic, musical and so forth. The psychological game, on the other hand, looked for the unity and harmony not so much according to choice, ordering, linking, chaining and confrontation of the contents, but more according to the meditation following each stage of the game onto which it put all its emphasis."[9]

Hesse's game can thus be visualized as a *game of interpretation*, as a game of interpreting the world within the questionable, observable worldly. Even if the world permanently presents itself to interpretation as one which is questionable: It is the concentrated and complete taking in sight which guarantees that interpretation is not being lost in the undeterminedly arbitrary:

---

[7] HESSE 1977, 37 sq.
[8] HESSE 1977, 130.
[9] HESSE 1977, 210.

„We shall direct the attention to everything, because everything can be interpreted.“[10] And this interpreting, as the main activity of the primarily hermeneutically constituted human, shows up as the translation of what has been perceived into a human universal language. The search for the latter is another ancient project of humankind: „[To study the whole game from the beginning to the end means] ... I work through each of its propositions, translate them from the game language back into its original language, into mathematics, into ornamentics, into Chinese, into Greek and so forth ... [W]e have invented and expanded the glass bead game in the run of centuries, as a universal language and method, in order to express all intellectual and artistic values and concepts and to bring them to a common measure.“[11]

With the formulation „to bring them to a common measure“ a universal human measure is actually being referred to. This is so because if it is anthropological qualities which are made the object of action in the glass bead game, what should have become clear by now, then we deal with the humans altogether indeed. And if humans are subject to what within the wordly are the actions of what they are able to perceive, then the problem is their own positioning with respect to that perceived, it is the uncovering of the relationship between person and world and between person and human collective. In other words: By means of inserting oneself into the world and by means of communicating about this with the others the sense of the game unfolds and offers itself to interpretation, at the same time. Essentially, the world communicates with itself and utilizes humans as an organ of communication. But the constitution of these humans is important, in the first place. Hence, the conception of the game reflects the picture of humans involved: „The human we mean and we would like to have, who to become is our objective, would be able to exchange every day his science or art with any of the others, in the glass bead game he would bring the most crystalline logic to light and within grammar the most creative phantasy.“[12] Hence, the game approaches a concrete utopia: „[The ideal], the thought of the interior unity of all intellectual efforts of humans, the thought of universality, has found its perfect expression in our noble game.“[13]

[10] Hesse 1977, 82.

[11] Hesse 1977, 125.

[12] Hesse 1977, 83.

[13] Hesse 1977, 249.

When we ask today what remains of this literary conception of Hesse's and what we can still utilize, then we discover in the anthropological centre of his ideas aiming at universality something which is still of importance, especially as far as the recent developments in the sciences and the arts are being concerned. Of course, we have to think of alterations in detail, because some of the concepts depend on the socialization of those who utilize them, and a straightforward translation is sometimes difficult, if not completely impossible. A central concept of this kind with Hesse is that of *meditation.* It should be difficult to find a satisfactory answer as to its universal meaning. Hence, we would prefer to actually replace this concept by that of *association* which appears to be more universal in semantic terms. This is especially so, because association is obviously a permanently present constituens of our everyday life, particularly in the sciences and arts, where it is condensed to an epistemic means of knowledge and understanding. Also, association introduces the aspect of the coincidental which appears to be relevant for human thinking, by being able to spontaneously shed light on an otherwise hidden sense. If asking what an actually playable version of the glass bead game could be today – presently a project under development indeed – then deterministic (logic, algorithmic, propositional) reflexion and pre-reflexive (non-propositional) association will be fixed ingredients.[14] The route towards a concrete version (possibly arranged as a computer game) is long however, although a suitable project is under way already.[15] The Tones & Notes project[16] is one step towards the objective. But it is obvious that such a project asks a lot of its participants, especially what individual openness of the mind is being concerned with a view to a multitude of fields involved and the readiness to accept system and method of the others. In other words: *It matters to assume nothing except the criteria of adequacy themselves.* Hence, the project is also an *ethical* project. In this sense, the appeal is to found many „forest cells“: „Gignit autem artificiosam lusorum gentem Cella Silvestris. (The forest cell [Waldzell] however produces the elaborate people of the glass bead players.)“[17]

---

[14] In the internet one can find a large number of groups who deal with the development of a playable version of the game. As far as we can recognize, all of these approaches are not very satisfactory until now.

[15] ZIMMERMANN 2004. For a first introduction see ZIMMERMANN 2003b.

[16] ANDREAS KUHN, CONSTANTIN CARL, RAINER E. ZIMMERMANN: FarbTöne. (Exhibition, Presentation & Discussion) Pasing (München), July 2003. See also my „manifesto“ for this occasion which is added here in the appendix.

[17] HESSE 1977, 82.

## *II*

In order to illustrate the heuristic power of a glass bead game of the kind described here, though visualized with respect to modern science alone for the time being, we choose an example from recent conceptualizations in physics, namely Smolin's *principle of cosmological natural selection*. Obviously, we cannot unfold the complete scope of all the interdisciplinary results which may be achieved by a conceptual game of the glass bead type, but what we would like to do here is simply to give a first impression of the basic problems involved. A more detailed example has been discussed in a recent book of mine.[18] I follow here the line of argument as displayed in a recent summarizing paper.[19]

### *1. Introduction*

Following the book publication by Lee Smolin[20], I have discussed his model of cosmological natural selection twice elsewhere[21], pointing to a number of inconsistencies of the model and possible variants which may be helpful in solving one or the other problem with it. After these preliminary discussions we are now capable of re-formulating Smolin's conception in a generalized way by means of applying directly the somewhat corrected version of the analogy between physics and biology actually being put forward. The advantage is to concretely establish *selection* as an intrinsically universal principle indeed. The Universe in the *facon de parler* of Smolin's is then nothing but the individual sample whose type is the analogue of the *phenotype*. Hence, we shift from the concept of selection to what we call *superselection* in order to point to the fact that the principle invoked here acts upon *networks* (populations) of Universes which have to be classified according to their type. The latter is being defined in the sense of the excess

[18] See note 15 above. – Note that in the meantime most of my Los Alamos archive papers have been collected in ZIMMERMANN 2002a.

[19] ZIMMERMANN 2003a (I utilize here a shortened und revised version of this preprint.)

[20] SMOLIN 1997.

[21] ZIMMERMANN 1999 – also under: http://www.ernst-bloch.net/akt/mitbei/klymene.zi. See also ZIMMERMANN 2003b.

productivity of black holes actually achieved, and is a measure therefore for the possible number of „offspring" of Universes which may be eventually produced. We have thus a population of Universes which can be stratified into sub-populations reflecting the various phenotypes. (The question as to whether a population may contain Universes which are not space-times, simply due to epistemological reasons, we leave aside for the time being.[22]) These phenotypes are involved in competition among each other. Superselection acts thus onto these phenotypes according to the environmental structure in which the sub-populations participate. Hence, what we need is the equivalent of an environment, some *intercosmic mediator*.

But let us first have a look onto the interior „ecology" of a *single* Universe: A generic Universe consists of a hierarchic structure of galaxies (supercluster, cluster, local groups, individual samples of galaxies). It is indeed the *stellar ecology* of galaxies which explicitly determines the production rate of black holes. Hence, also a galaxy can be visualized as an analogue to local ecosystems containing populations of phenotypes. In this case the types are *stellar* types plus a mediator which is given in terms of the *interstellar matter*. The Universe appears thus as a hierarchically organized population of populations of local ecosystems. But if the stellar type (described by the spectral class of stars) is a local version of a phenotype, what is then the associated genotype?

Remember that biology is essentially organized in terms of three structural and functional levels which we can call the molecular, cellular, and organismic levels, respectively. The last one is directly associated with the phenotype. The molecular level coincides with the level for which this expression has been introduced in the first place, namely the chemical level. So what is the cellular level? This would be exactly that level on which the „cosmogenetic" coding is being stored, in a place which would be the equivalent of a *cell nucleus*.

A short intermediate remark to the utilization of Smolin's selection metaphor: Indeed, at the occasion of a short discussion of the topic some time ago[23], Smolin has not shown much enthusiasm when being confronted with the arguments unfolded above. Instead, he seemed to be pointing to the fact that this metaphor chosen by him would only be applicable in a

---

[22] ZIMMERMANN 2000. See also ZIMMERMANN 2002b, 165-186.

[23] Discussion with Lee Smolin at the occasion of the conference „Key Tests for Cosmogenic Theories", Newton Institute, Cambridge (UK), 6.-9.12.1999.

somewhat limited way and could not directly be transferred to a physical terminology. But here we have to object that the metaphor is apperently of *some* use after all. However, this being the case, it has to be developed with all its consequences. And the latter are to be examined carefully with a view to their consistency. If they turn out to be inconsistent, the principle has to be given up. But there cannot be a mere *semi-metaphor* in sufficiently general terms. Hence, a complete discussion of the principle introduced is what is being asked for.

## *2. A First Ansatz*

Hence, what we do is to start in a straightforward manner from the original analogy which is itself referring to the organismic structure of organization as known from biology: Therefore, phenotypes are types of organisms which are themselves structured in a molecular way and constituted on that molecular level by means of the respective genotype. It is important here to remember oncemore that we will not expect that we actually deal with explicitly biological quantities after all, but that instead we will have to deal with physical quantities only. Insofar the metaphor stays nothing but a metaphor. On the other hand, we know of course, that biological systems are again nothing but chemical and thus complex physical systems themselves. Hence, we can legitimately argue to found the further analogy onto the relative orders of magnitudes being actually involved: The individual body of human beings e.g. represents a sample of the human phenotype and thus constitutes the organism on the macroscopic level in terms of biological aspects. Let us say that the human body contains some $10^2$ organs, which are composed of different sorts of webs (according to the various fields of sensory perceptions, food processing, motion, reproduction, and so forth). With roughly $10^{14}$ cells per organism this comes to about $10^{12}$ cells per organ. (The molecular mass of human cells is at about $10^{15}$.) Hence, the ratio of the „highest" to the „lowest" level of organization is about $10^{10}$ or $10^{12}$, according to whether one refers to the organs or the whole organism.

On the other hand, the Universe contains about $10^{11}$ galaxies of which each for itself contains $10^{11}$ stars on the average. The galaxies are organized in clusters and superclusters with about $10^6$ and $10^9$ galaxies, respectively.

Hence, it is intuitively clear, within the chosen analogy, to actually identify the level of superclusters with the level of organs, and the level of galaxies with the level of cells. The „scalelessly" mediated orders of magnitude among clusters refer then to the levels of different biological webs. Note that in biology however, organs are usually not only classified according to their *structure*, but also according to their *function*. That is, organs have a specific function within the whole organism. Moreover, there must be a *fine tuning* of functions among each other, because any malfunction has to be compensated for a while for not endangering the organism altogether. In the long run, any malfunction will put the organism at a decisive risk. Nothing else is the case with the physical functions of the constituents of the Universe.

If now galaxies are the equivalent of cells, what would be formally equivalent to the nucleus of a cell? The point is that the biological analogy can be carried quite far in the case of galaxies, because like cells they indeed perform specific functions of some „metabolism" and can therefore be visualized as regulatory systems. (This is in fact what we also claim when speaking about body cells of an organism.) The sort of metabolism mentioned here is however one of a cosmic exchange of matter, where the important things happen in places we cannot observe very well: in the dark spaces in between spiral arms of galaxies. Because this is where stars are being „cooked" which „cook" themselves other matter then and so forth. It is from then on that they become visible to us. Hence, the complete galaxy gains the connotation of a self-organizing (massively parallel) computational system, very much like the biological cell.

The cells utilize the information which is contained in the genetic structure and is actually stored in the cell's nucleus, and produce proteins from it. The DNA is very much like a hard disk, while the proteins are a sort of RAM.[24] We can claim something similar with respect to the galaxy, provided we think of its „nucleus" as of a comparable system for storing and processing information. Indeed, there is a literal nucleus to most galaxies which is represented by a *central massive black hole*. In other words: It is such a black hole which would be the adequate place to „store the laws of nature in some symbolical manner" as Smolin is himself asking for. (He actually disputes the possibility of such a storing of symbolical information though. But that is a point where doubts should be in order.) Obviously, the next ques-

[24] CF. SIEGFRIED 2000. Note in particular chapter 5: The Computational Cell (95-113).

tion is *how* the relevant information could be stored in black holes. Unfortunately, the interior of black holes is not well-known until now.[25] However, massive black holes in the centres of galaxies as they have gained renewed relevance since the beginning of 2000 when the Chandra satellite observatory did its first impressive series of X-ray photos, could well be visualized as that „gates towards a new physics" as Martin Rees has introduced them recently.[26] The characteristic „careers" of black holes have been studied in detail some time ago by Kip Thorne among others.[27] Many aspects recovered here correspond nicely to what Smolin proposes in his approach.

These aspects by the way imply also the absence of the fundamental categories of space and time which we have discussed elsewhere.[28] Thorne e.g. compares them within the context of a black hole with a piece of wood soaked with water coming into a fire. Time (the water) evaporates and the remaining space (dry wood) becomes ashes (Wheeler's quantum foam).[29] And indeed, within this physics of the quantum foam we should look for the analogy of (cosmic) nucleotides. Not in the foam itself (because the storing of information should not be subject to random fluctuations), but as its first product. The latter might turn out as *spin networks* in fact.[30] Hence, the clarifying of the analogy is closely related to the problem of developing a TOE.

The same is actually true for what we can visualize as „intercellulary" space: The respective region in between the galaxies carrying the reserves for the latter's metabolistic functions is typically visualized as an almost perfect void. But even if this should be true, there is still gravitation. Also within galaxies, the patterns of interstellar matter recently observed could be understood as resulting from exterior „environmental" inputs establishing explicit forms of harmony.[31]

---

[25] A long time ago I have worked myself on the inside story of black holes before changing into other fields. In the meantime, I noticed that presently, there is (as far as I can see) only one group engaged in this topic, namely that of Werner Israel in Canada (and its former members elsewhere in North America, to be more precise). And the insight gained recently is not very much impressing, to say the least.

[26] REES 1997. In particular chapter 5: Black Holes: Gateways to New Physics.

[27] THORNE 1994 [The Einstein Legacy. Quoted here according to the German edition.], 524, 545. As to wormholes cf. 556.

[28] See also more recently ZIMMERMANN 2001b.

[29] THORNE 1994, 544.

[30] ZIMMERMANN 2001a.

[31] SMOLIN 1997, 132. – Cf. more recently: NATURE 394 (1998), 524 (when S. Battersby is talking on B. G. Elmegreen et al. in AP.J.LETT. 503 (1998), 119-122).

## *3. Computational Aspects*

But there is still another point to that: The internal structure of the Universe alone cannot be decisive for superselection, because we deal with *populations* of Universes in the first place. That we can retrace the basic structure of selection also within the individual sample and all of its substructures is only an epiphenomenon of the underlying model's scale independence (which is actually a necessary condition for our evolutionary principle in order to adapt it to self-organized criticality in the sense of the Santa Fe school). Hence, we have to differ between *internal* selection in the sense of Smolin (within a given Universe subject to its stellar ecology) on the one hand, and *external* selection of types of Universes (superselection) on the other hand. Obviously, the one has to be fitted to the other. In fact, what we can do is actually to visualize internal selection as a projection (more precisely: a projected image) of superselection. In other words: The biological selection proper shows up then as a mere differentiation of superselection projected onto the planetary ecology. This would indeed shed some light on the question of possible earth-like planets.

But if so, the remaining question is for the „cosmic mediator" already mentioned. It would define the background of Universes on which populations unfold. Hence, it could be visualized as that environment whose structure gives the initial drive for the competition of phenotypes in the first place, because it is only a *finite* environment, one with *restricted* ressources, in other words: a *physically deficient* environment, on which selection can be sufficiently founded. It may be worthwile to look for this mediating background (which is not a geometrical background as we know it, because we are talking about a region which is *beyond* Universes, and thus beyond space and time) in the loops of loop quantum gravity itself. In a first instance, the concept of spin networks is relevant here. The problem is that we have to permanently think „in terms of space and time" so that we cannot abstract from these fundamental categories, even if talking about their factual absence. As I have discussed elsewhere[32], spin networks can be visualized as the *boundary* of space-time, and as such they are also the (epistemological) boundary of substance (in some modern sense).[33] But the point is that there is no real transition from the one (the world) to the oth-

---

[32] Cf. note 28 here.

[33] Cf. notes 19, 20 here.

er (substance) and viceversa across that boundary, because substance *is always everywhere* (thus non-locally) underlying the world which we can observe, but which is nothing but a back-projection of substance (which we cannot observe). The question is whether seen under this ontological perspective, it is useful to think of physics as the foundation of biology all the time, while not thinking of the viceversa: that also biology would have something to contribute to the foundation of physics in turn. This may be so because after all, physical theories are being produced by biological living beings according to what the latter can actually perceive. This perception however, together with the thinking applied to it, is primarily biologically constituted. Recently, Louis Kauffman has oncemore reminded on this point and tried to describe a relationship between biology and logic which might be of a specific systematical meaning for what we have said here.[34] Again, there is another parallel to the conception of visualizing the Universe altogether as an emergent computational system which in the case of biology differs from the computers as we know them only in the fact that it is software and hardware at the same time.

## *4. Black Holes*

But the essential problem with black holes is as to their eventually becoming „quantum“: Originally, the programme was quite straighforward. If a star with a certain minimal mass underwent gravitational collapse, then sooner or later a horizon would form defining the black hole‘s boundary. And the work concentrated on describing horizon properties. The actual vicinity of the singularity was then cut out from both the space-time manifold and from the discussion, respectively. That was what we learned as students when reading one of our three bibles at the time.[35] Things became more complicated however when Hawking, Bekenstein and others started to discuss quantum aspects of black holes. From the beginning on, the suspicion emerged that classical concepts would be carried over to quantum situations without being really justified. Not that this would have been something new when dealing with quantum physics (indeed it actually es-

---

[34] KAUFFMAN 2002.

[35] Namely the books of HAWKING and ELLIS, of MISNER, THORNE, and WHEELER, and of WEINBERG, respectively.

tablished a great deal of distinction for those who worked in „classical" field theories when visualizing the whole Schroedinger programme as some kind of elaborate guessing according to concepts of classical physics). But in the case of black holes it seemed somehow to go too far. All of this despite some very attractive aspects of the new conceptions such as Bekenstein's idea of treating black holes as an analogue to atoms in pre-quantum physics and so forth. However, with the advent of *black hole thermodynamics* and in particular with the treating of the horizon as a material membrane in the sense of Thorne, things became very interesting, and one was somewhat distracted from the quantum problems lurking underneath. As it turns out now, the very terminology utilized when talking about quantum black holes is covering most of the problems which still have remained. Take the example of the celebrated *Planck mass* which serves as a criterion of talking about a domain where all relevant forces actually meet[36]: It is the frequent changing of the systems of units involved what is diverting from the fact that it is not the Planck mass which is relevant at the Planck level, but instead it is the dimensions involved which are relevant. In other words: If you have a piece of matter with roughly $10^{-8}$ kg of mass, then you do not have normally any problem with quantum fields or quantum gravity as to that – because it is only when you compress this mass to the small Planck length dimensions that it would become theoretically relevant. So it is the *volume* rather than the mass which is important after all. Obviously, when dealing with stellar black holes, you are not dealing with any Planck mass whatsoever, by the very definition above. In fact, we are talking of several solar masses. And the mass remains conserved. What is changing while the star performs a gravitational collapse, is the mass *density* which is enormously increasing because the available volume is decreasing. (And this tells you something about the characteristic gravitation involved which points towards tidal forces rather than to anything else – which also tells you something about the entropy change being involved and explains why a black hole is *not* really a time-inverted white hole.) We know in the meantime from the seminal paper of Rovelli and Smolin and from other papers dealing with its consequences[37] that the volume (as well as the area) of space has to be visualized as a discrete and thus

[36] BAEZ 2001, 177-195.

[37] ROVELLI & SMOLIN 1994, and LEWANDOWSKI 1996, as well as ASHTEKAR & LEWANDOWSKI 1996; 1997.

quantized entity. In fact, in case of the volume the quantization is not quite as straightforward as in the case of the area, but we notice that the spectrum runs proportional to $l_P^3$ as should be expected. This insight has resulted in a number of consequences in the field of quantum information lately.[38] Hence, with respect to what we would like to have in the case of black holes when thinking of what we discussed above, the explicit coupling of gravitation, thermodynamics, and quantum information is very promising after all. Alas, there remains the unsatisfactory situation that the transition of black holes from the classical into the quantum domain is far from clear. The quantization of area and volume of space mentioned above gives some operational rules as to dealing with the computation of quantities at the Planck level. But the important problem here is the following: While undergoing collapse *each* black hole will *necessarily* come into the quantum domain by means of its permanently shrinking volume. And then the question will be how collapse is actually being *halted* at the Planck length, because the true zero-point of length cannot be reached due to the convention of declaring the Planck length to be the minimum length which could be possibly attained. Hence, we would conclude that there are *no objects* with a length which is smaller than the Planck length. But note that a black hole arriving at the Planck length by means of gravitational collapse will not have any Planck mass then! In fact, the mass will be much larger than that which means that the Compton length is actually *undefined*, because it would be smaller than the Planck length for such cases. In other words: While the black hole is entering quantum dimensions, it does *not behave* like a quantum particle at all. If we cannot ascribe a Compton length to it, then it is not subject to Heisenberg's uncertainty relationship!

In fact, it is quite clear why this should be the case: The reason for this is that within the region of the Compton length, the length scale goes as 1/m, while for the macroscopic region (where the Schwazschild length is of relevance) it goes as m. This is indeed due to the onset of Heisenberg's uncertainty relation. But the question remains why collapsing black holes do not seem to be subject to this transition. On the other hand, postulating *virtual* black holes which are being created by means of spontaneous fluctuations out of the vacuum does not really help here, because then the question remains whether it is legitimate at all to call such objects black

[38] ZIZZI 1999 (also in Entropy 2 (2000), 39 sqq.); 2000 (also in J. GRG 33 (2001), 1305 sqq.); 2001a; 2001b.

holes, if they have not undergone any gravitational collapse before. As it turns out, there are basically two desiderata: One of explicitly demonstrating the transition from a classical volume to a quantum volume displaying the relevant effects such a transition would have for an object undergoing gravitational collapse, and another one of explicitly studying the relationship of length scales at the Planck scale proper as well as defining the state of an object whose Compton length is smaller than its Schwarzschild length, in order to be able to decide whether this could really be called a black hole or not. A first step towards approaching this problem is a recent paper of Paola Zizzi[39] . Further work with more technical explications is in progress.[40] Hence, it is likely to find the information processing mechanism discussed above within the interior of black holes. In the end it is this mechanism on which the concept of cosmological selection is actually founded. But it does not suffice to elaborate further on quantum mechanical details of black holes without being able to say something about the transition from classical to quantum states during gravitational collapse and without clarifying the nature of black-hole like objects at the Planck scale in more detail.

## *6. Some Conceptual Conclusions*

Finally, how to come back to the glass bead game? And what about the interdisciplinary perspective discussed earlier? Note that the leading question of the example discussed in the preceeding is that for the relationship between the world and its foundation. Human research activities aim at the modeling of the world. But the world is not enough. There is also its foundation. And in fact, as it turns out, the world as we model it shows up as the foundation's self-recursion. In other words: The categories of space, time, and matter are necessary and sufficient to model both – world and foundation. Although the latter is by definition outside the scope of the mentioned categories. Hence, humans do model the world according to their linguistic capacity, and this in turn is determined by their perceptive and thus cognitive capacity. This is also at the very root of the problem when dealing with populations of Universes as in the case of our chosen

---

[39] ZIZZI 2003.

[40] ZIZZI & ZIMMERMANN: Virtuality & Actuality of Black Holes (forthcoming).

example. But from this conceptual starting point we derive further, secondary, problems one of which is that of the universality of a principle (like the principle of natural selection), which reflects the underlying relationship between local and global, and between singular and universal, at the same time. Note also that there is an intrinsic search for harmony included in all these partial approaches to the totality of what we call world. Hence, we are explicitly dealing with the quest for unity within the manifold multitude we actually observe. But, even in conceptual terms, we will never leave the world *as it is perceived*, and we will never grasp the world *as it actually is*. In fact, all these aspects are common to all the various fields involved. This is the reason for being able to display analogies among the sciences and arts, without leaving the conceptual framework of human reflexion. This framework however is essentially founded on the mediation of knowledge and action, hence of epistemology and ethics: Physics models the producing ground of the world and the physical production of logic, at the same time. Hermeneutic is itself logically constituted, because it has a rational nucleus (in the sense of the Leibnizian *nihil sine ratione*), and so is the semantic of signs which can be derived from it in semiological terms. Hence, even aesthetics cannot be separated from physics at all, and, finally, ethics is essentially knowledge of all of this. The actual production of that knowledge progresses in the manner of the glass bead game: The playing board is the local framework of space, time, and matter, the rules are given by logic (but as this is produced itself by physics, the rules of logic directly reflect the laws of nature), the pieces of the game are propositions to be placed into their relational network called theory, and adequate (i.e. ethical) actions are at its stake. In fact, the *form* (Gestalt) of this relational network expresses the theoretical *contents* of what the propositions actually tell. This is what we mean with harmony: the fine tuning of form and contents.[41] But it is a game after all, we never deal with the „real thing"

[41] This can be clearly demonstrated when breaking the preceeding text on cosmological natural selection up into single propositions and re-arrange them within a relational network of the conceptual context. Imagine them ordered in a three-dimensional lattice with the types of propositions being labelled by appropriate colours. What we get then is the coloured „cloud" of a theory exhibiting a specific shape in our representational 3-space. The idea is then that the form corresponds to the self-consistency of the propositions involved such that we can utilize the form as a macroscopically observable criterion for that consistency. After a while we would be able to deduce from a „good" or „bad" form whether the underlying theoretical contents is useful or not – very similar to the intuitive association generated by the (two-dimensional) shape of

as such. This can be clearly seen in case of what we call „laws of nature", because we refer to them as they are modeled by ourselves, but not as they actually are. The very concept of law itself is nothing but a mapping. Hence, the original definition of games according to Schiller is still valid: To be more precise, in that famous passage, Schiller speaks of the human driving force, the *impulse*, to play and calls it the connection of the impulse to change in order to give time a contents, and of the impulse of form in order to actually negate change and sublate time. Hence, the intention of the game is *to sublate time within time*, and to reconcile being with becoming.[42] This demonstrates clearly the relationship between physics and ethics. So Schiller concludes: „Hence, the impulse to play, in which both of them [the two impulses] act in a connected manner, will compel the mind in moral and physical terms, at the same time; hence, it will also sublate all of the compulsion itself, because it sublates all contingency, and thus sets humans in liberty [releases them to freedom], physically as well as morally."[43]

Shortly afterwards, Franz von Baader, a friend of Schelling's, published his paper on the „founding of ethics by physics"[44] advocating a concept of nature which visualizes nature as a composing process activity which is structured in itself in a hierarchical manner.[45] Although his approach is definitely one of idealistic, if not even mystical, connotations, its general idea, namely that of a permanently emerging nature, an active *natura naturans* and in the sense of what he chose as first principle of physics: *Ignis ubique latet.*[46] is still of significant relevance for us today. It is in fact latency which underlies human reflexion, and it is reflexion which eventually produces a tendency out of this latency by means of emerging concepts. This is after all what the glass bead game is all about. So Baader's principle can be actually completed by what the materialistic philosopher Ernst Bloch once formulated referring to his own perspective of a somewhat dif-

---

the positions of pieces in a Go game. This is indeed a straightforward means of incorporating intuitive association within the reflexive (propositional) thinking of theory production. This conception is explicated in some detail in ZIMMERMANN 2003b.

42 SCHILLER 1989, 612 sq.

43 SCHILLER 1989, 613.

44 VON BAADER 1921.

45 VON BAADER 1921, 7, 9.

46 VON BAADER, 12.

ferent context: *Quidquid latet, apparebit.*[47] Hence, the worldly process of modeling the world shows up as an experiment the world is performing utilizing itself, and it includes its own recording and produces thus its own narrative. It is the world's foundation though which is playing a game with itself which in turn is the conditional space of free play for the world. The inventory the foundation utilizes has been coined recently SOAPS [ = Space Of All Possible States] by Devin Harris[48] so that we could visualize the *experimentum mundi* (in the sense of Bloch) as an actualization of „SOAPS bubbles".

## *Acknowledgements*

For helpful suggestions and illuminating discussions I thank Constantin Carl, Christian Fuchs, Inga Gammel, Wolfgang Hofkirchner, Franziska Holzner, Tarja Kallio-Tamminen, Michael Keller, Gerd Koch, Andreas Kuhn, Sabine Ley, Annette Schlemm, Heike Siegert, Doris Zeilinger, Paola Zizzi.

## *Bibliography*

ASHTEKAR, ABHAY & LEWANDOWSKI, JERZY

1996 "Quantum Theory of Gravity I: Area Operators" in http://www.arXiv.org/pdf/gr-qc/9602046.

1997 "Quantum Theory of Geometry II: Volume Operators" in http://www.arXiv.org/pdf/gr-qc/9711031.

BAADER, FRANZ VON

1921 "Über die Begründung der Ethik durch die Physik" in Pulver, Max (ed.): *Schriften Franz von Baaders*. Insel: Leipzig. 1-30.

BAEZ, JOHN C.

2001 "Higher-dimensional algebra and Planck-scale physics" in Callender, C. & Huggett, N. (eds.): *Physics Meets Philosophy at the Planck Length*. Cambridge: Cambridge University Press. 177-195.

---

47 BLOCH 1975, 148.

48 HARRIS 2003.

BLOCH, ERNST
1975 *Experimentum Mundi.* Frankfurt a.M.: Suhrkamp.

HARRIS, DEVIN
2003 *Integrating Absolute Zero into the Set of all Possible States.* Preprint, August 2003.

HESSE, HERMANN
1943 *Das Glasperlenspiel.* Zürich: Fretz & Wasmuth.
1977 *Das Glasperlenspiel.* Frankfurt a.M.: Suhrkamp.

KAUFFMAN, LOUIS H.
2002 "Biologic" in http://www.arXiv.org/pdf/quant-ph/0204007 v2.

LEM, STANISLAV
1983 *Solaris.* München: DTV.

LEWANDOWSKI, JERZY
1996 "Volume and Quantizations" in http://www.arXiv.org/pdf/gr-qc/9602035.

REES, MARTIN
1997 *Before the Beginning. Our Universe and Others.* London: Simon & Schuster.

ROVELLI, CARLO & SMOLIN, LEE
1994 "Discreteness of Area and Volume in Quantum Gravity" in http://www.arXiv.org/pdf/gr-qc/9411005.

SCHILLER, FRIEDRICH
1989 *Über die ästhetische Erziehung des Menschen in einer Reihe von Briefen* (14. Brief) in Schiller, Friedrich: *Sämtliche Werke*, Band 5. 8. Auflage. Darmstadt: Wissenschaftliche Buchgesellschaft.

SIEGFRIED, TOM
2000 *The Bit and the Pendulum. From Quantum Computing to M Theory – The New Physics of Information.* New York: Wiley.

SMOLIN, LEE
1997 *The Life of the Cosmos.* Oxford: Oxford University Press.

THORNE, KIP S.
1994 *Gekrümmter Raum und verborgene Zeit.* München: Droemer-Knaur.

ZIMMERMANN, RAINER E.
1999 *The Klymene Principle.* Kassel: Kassel University Press.
2000 "Loops and Knots as Topoi of Substance. Spinoza Revisited" in http://www.arXiv.org/pdf/gr-qc/0004077 v2.

2001a "Recent Conceptual Consequences of Loop Quantum Gravity. Part I: Foundational Aspects" in http://www.arXiv.org/pdf/physics/0107061.
2001b "Recent Conceptual Consequences of Loop Quantum Gravity. Part III: Postscript On Time" in http://www.arXiv.org/pdf/physics/0108026.
2002a *The Physics of Logic*. Kassel: Kassel Unversity Press.
2002b "Spinoza in Context: A Holistic Approach in Modern Terms" in Martikainen, Eeva (ed.): *Infinity, Causality and Determinism. Cosmological Enterprises and their Preconditions*. Frankfurt a.M.: Peter Lang.
2003a "Cosmological Natural Selection Revisited. Some Remarks on the Conceptual Conundrum and Possible Alleys" in http://www.arXiv.org/pdf/physics/0304053.
2003b *System des transzendentalen Materialismus. Neue Anleitung zum Glasperlenspiel.* Paderborn: Mentis (in press).
2004 *Topos der Materie*. Paderborn: Mentis (in preparation).

ZIZZI, PAOLA A.
1999 "Holography, Quantum Geometry, and Quantum Information Theory" in http://www.arXiv.org/pdf/gr-qc/9907063 (also in *Entropy* 2 (2000), 39 sqq.).
2000 "Quantum Computation Toward Quantum Theory" in http://www.arXiv.org/pdf/gr-qc/0008049 (also in *J. GRG* 33 (2001), 1305 sqq.).
2001a "The Early Universe as a Quantum Growing Network" in http://www.arXiv.org/pdf/gr-qc/0103002.
2001b "Ultimate Internets" in http://www.arXiv.org/pdf/gr-qc/0110122.
2003 "Spacetime at the Planck Scale: The Quantum Computer View" in http://www.arXiv.org/pdf/gr-qc/0304032.

## *Appendix*

Recte

### *Tones & Notes*

Towards a New Fundamental Form of Expressive Representation.
A Manifesto.

The world is being perceived by humans under a special perspective only, in a two-fold manner: on the one hand in the sense that the human viewpoint is generically restricted to what is available to observation – due to biological and anthropological reasons – insofar as humans can merely perceive fragments of the world but not the world as it really is, i.e. as it is independent of their perception. But this perception refers exclusively to what is in the senses, hence what is physical by nature. So human reality is always restricted altogether. On the other hand it is also restricted in the sense that additionally, each person perceives the world under an individual, specific perspective. And the communicative clarification of what has been perceived, which is the condition of intersubjectivity, can only be achieved, if this singular perspective is permanently being re-processed with respect to all other such perspectives within a process of building a (discursive) consensus. Intersubjectivity thus becomes objectivity and constitutes a governing picture of the world. The two large discursive regions of worldly constitution, the sciences and the arts, are being organized therefore – not the least due to purely pragmatic reasons – in different and clearly differentiated sectors of design which refer to each other like mutually irreducible regions. But they simply represent specific and thus perspectively characterized fragments of the world which are being perceived by humans and are being modeled according to the perceptions actually made, and re-constructed in a more or less provisorical and thus transitory manner. *In other words, the world as it is, is only one, but the world as it is perceived, can be represented and expressed by many.* Hence, there is an obvious multitude of the world which is nothing but the mapping of the unity of the world. It is straightforward to suppose that the systematic synthesis of the various aspects of that same underlying One will be able to shed some light onto the interactions of these aspects among each other and thus onto the totality of which they are aspects. Hence, how the practical

*Verso*

interdisciplinarity in the sciences has produced new insight into various sections of reasearch recently, a likewise interdisciplinary approach to the arts can add to a widening of the epistemic horizon and an enlightenment of the relevant context. It is in particular the conceptual though distanced confrontation of otherwise irreducible forms of art which initiate the unfolding of an associative space of free play for those interactions in whose framework traditional categories of viewing the world can be sublated and innovative experiences can emerge. As far as the design of fine arts, e.g. in the case of paintings, works primarily with the category of space, while music works with the category of time, both of them however encompass the complementary category, respectively, it is straightforward to investigate the practical co-operation of both within their space of free play which is spanned by the interaction of these categories. This in turn offers itself to perception in opening up a new combinatoric of harmonical forms which enrich the multitude of worldly forms even more. Essentially, we deal with a topological and geometrical combinatorics which is subject to its own laws of form such that relevant characteristica may be expressed by means of similar structures of mappings: e.g. by means of symmetry principles or symmetry breakings, the modulation of continuous motion, the spontaneous succession of evolutionary jumps and so forth. Hence, the composition of Tones & Notes aims to a reduction of complexity, in the first place. This is so because humans though disposed of selecting available information all the time, for getting rid of irrelevant information, they are exposed, especially within the cultural sector, to an avalanche-like abundance of information which they hardly can cope with anymore. Hence, it appears to be useful to go back onto the fundamental forms of perception and to take the result as a basis for re-assuring itself about the approach to the world. So as far as paintings are concerned, elementary shapes and basic colours and tones shall be in the focus of attention, while as far as music is concerned, the optical impressions shall be complemented by spontaneous compositions (improvisations) in the fundamental notes.

# Philosophical Consequences of the Gödel Theorem

Alfred Driessen

## 1. Introduction

Looking at the program of this symposium and reading the title of the talks could easily provoke certain astonishment: why should one bring all these things together? What has physics, mathematics in common with philosophy or even theology? Are we not mixing disciplines which should not be confused and ending up in a big mess without scientific value. To make this point more clear, why should or even principally could mathematics change a pure philosophical argument?

In order to answer these questions one could start by observing the human desire to bring all knowledge about reality together in a single, consistent view. It is the dream of the philosophers of all times to create such a view, a consistent metaphysics without internal contradiction and in accordance with experience. It is based on the assumption that reality is by definition consistent, complete, without contradiction and accessible to the intellect.

Already in the times of Aristotle the possibility of a sound metaphysics was rejected by the skepticism of sophists, who denied the possibility of finding the truth. Aristotle argued that the fundamental statement of philosophical skepticism, that it is not possible to affirm anything truly, is already an absolute statement not demonstrable by science. By accepting this statement as an absolute truth, one is therefore in contradiction with its meaning[1]. The search for a unifying metaphysical view, therefore, is challenging, but in my opinion the only one in accordance to human intellectual dignity.

[1] See Aristotle, *Metaphysics*, Book XI - Part 5

Reflecting on the huge amount of information provided by his senses man is confronted with a very fundamental question: is there a reality that can not be reached by senses even not with the aid of sophisticated instruments? Is there a reality, which does not obey the physical laws of nature? This is certainly a philosophical question that is answered differently in different metaphysical systems, but as will be shown below, certain experiences support or make difficult certain metaphysical systems. For example, Aristotle based his metaphysics on the everyday experience of the occurrence of real changes. In the philosophy of Parmenides these changes were considered to be only apparent and not based on reality. His philosophy, therefore should and has in fact be disregarded.

Today the empirical facts have largely been increased, and, what is probably more important, these facts are only accessible to a small number of well-prepared specialists. How can a philosopher understand these new empirical facts – often only expressed in the highly formalized language of physicists, mathematicians or other scientists - and evaluate their relevance with regard to his metaphysics? Not to speak about a scientific discussion on the validity of these facts.

In the following an attempt is made to analyze an important mathematical discovery, the theorem of Gödel and to demonstrate that the acceptance of this theorem has a direct impact on the consistency of metaphysical systems. It will be shown that there are elements of reality, in this case the human intellect, that can perform actions that exceed the potential of operators exclusively subjected to physical laws.

To make it very clear, the Gödel theorem is in the field of pure mathematics. The philosopher as philosopher can not decide on the rightness of this theorem. What he, instead can do, is evaluating the general acceptance of this mathematical finding and reflect on the consistency between consequences of the mathematical theorem with consequences of his metaphysical view.

In the following the contribution of three mathematicians will be presented: first Gödel[2] himself, then the further elaboration by Turing[3] and finally the consequences for the human mind as worked out by Penrose[4]. Schins[5] ,[6] presented in several contributions the basic ideas of these authors.

[2] GÖDEL 1931, 173.

[3] TURING 1936-1937, 42, 230-265; 43, 544-546.

[4] PENROSE 1989.

[5] SCHINS 1997, 49-56.

[6] SCHINS 2000.

In this paper often use will be made of his argumentations. Finally we will present a philosophical evaluation.

## 2. The Gödel incompleteness theorem

In order to grasp the impact of the work of Gödel one has to consider the foundation of mathematics at the beginning of the 20th century. There was a strong school considering mathematics just a product of human logic. In this view it is principally possible to arrive at any mathematical truth by applying exclusively a set of axioms and formal deduction. The German mathematician David Hilbert expresses this conviction by: *In Mathematics there is no Ignorabimus*[7]. He formulated in 1900 the 10 last real problems of mathematics to be solved. The last of these, the *Entscheidungsproblem*, concerned the existence of an algorithm able to solve general classes of mathematical problems[8]. The solution of this problem would be the triumph of the formalistic approach: just following an algorithm, i.e. a sequence of deductive steps one would arrive on the solution. The question of meaning would be irrelevant; as long as the formal procedures would be used one would arrive at the mathematical truth.

In his communication of 1931, Gödel[2], an Austrian mathematician, demonstrated the fundamental inadequacy of the pure formal approach. He proved that in every consistent, sufficiently general axiomatic system (i.e., systems which are based on axioms and specific rules of deduction)

- there always exists a true proposition which cannot be deduced from the axioms (Gödel's incompleteness theorem);
- the consistency of the axioms cannot be deduced from the axioms (Gödel's consistency theorem).

In both theorems Gödel states that a certain mathematical truth cannot be obtained exclusively by deduction, i.e. exclusively by a formal approach. Mathematical meaning and mathematical truth are relevant concepts not reducible to formal logic. To prove his incompleteness theorem, Gödel

[7] HILBERT 1935, 297-298.
[8] HILBERT 1900.

provides a proposition whose correctness and truth everyone with a minimum of mathematical education can confirm, but which is undecidable within the formal system in question. It is here not the place to follow in detail the proof of Gödel; it is, however, worthwhile to express in words the so-called Gödel proposition in the incompleteness theorem: *there exists no proof for the Gödel proposition.*

In order to get a taste of Gödel's argumentation one can start with the observation that the Gödel proposition is legitimate (as is demonstrated formally by Gödel). There are now two alternatives, there is a proof for this proposition or not. If there is a proof, a legitimate proposition that is formulated according the axioms of the formal system is wrong. Consequently there is a contradiction within the axiomatic system which is impossible and the assumption has to be rejected. If there is no proof, the proposition is true and the incompleteness theorem is correct. This is the only alternative without formal contradiction. One is therefore forced to accept the validity of a proposition which, being true, is undecidable and cannot be proven in the formal system.

It is not so easy to realize the impact of the work of Gödel. What directly is shown is that mathematics goes beyond applying deductive or formal steps, also insight, intuition or meaning on a higher level come into play. Further, no formal system is complete, as a mathematician can always provide true statements that are formally improvable.

A first important conclusion can already be drawn now. For some physicist, the reduction of physical reality to a single theory or even a single formula would be the ultimate triumph of science. A physical theory, however, or a formula are part of a formal system, that after Gödel we know to be necessarily incomplete. Reality and also physical reality is by definition complete. The Grand Unified Theory (GUT) and other ultimate approaches will be necessarily incomplete and inadequate to describe the full richness of physical reality.

## *3. The contribution of Turing*

In 1936 Turing, a mathematician inspired by Gödel, conceived the ultimate computer, the Turing machine[3]. We are used to the fact that every three years we have a new computer with largely improved capacity. The Turing machine is the end of any possible evolution with regard to software as well as hardware: with infinite speed and memory, using digital logic, neural networks, quantum logic or any other up to now known or unknown technology for data processing. It is the best what physical laws and the most advanced design can offer with in addition unlimited speed and unlimited memory. With this machine any finite sequence of processing steps could be carried out in an infinitesimally small amount of time.

Turing posed himself the question and gave an answer with regard to the halting-problem. Is there a general algorithm that can predict whether the Turing machine will stop with a given program and given input? If that would be possible, one could look for certain prepositions and demonstrate, whether they are of the Gödel-type or not. Once these have been identified, one could isolate them and could continue with a 'clean' formal system. Turing showed that this algorithm does not exist, the algorithm that should mark Gödel-type propositions would be caught in an infinite loop in the Turing machine.

What can one learn from Turing?[9] Scientists and fiction writers are speaking about intelligence in man-made apparatus. There is a clear evolution: increasingly processing speed, memory and code complexity is introduced. But even after evolution and many centuries of future research there will be mathematical problems, these artificial intelligence (AI) devices can not solve. For the Turing machine, which exceeds these AI devices by far, has been shown to never come up with a result on well-defined algorithmic problems.

The question remains, would it not be fantastic if human beings would have the capacity of the Turing machine. Would this be a step forward or something destroying humanity in its very root? To answer this question one should not forget that the Turing machine is the best matter or beings exclusively bound to physical laws can deliver.

## *4. Penrose and the human mind*

In his book *The Emperor's new mind* Penrose[4] uses the results of Gödel and Turing to analyze the cognitive activity of the human mind. His argumentation is based on meta-mathematics, a level achievable by mathematicians but not exclusively bound to formal argumentation. His argumentation starts

---

[9] CACACE 1997, 27-40.

with two observations well within the field of mathematics.

There are mathematical results on the truth of certain prepositions, which can be recognized by any mathematician, but can not formally be proven (for example the Gödel preposition).

In addition, he learns from Turing, that the ideal computing device, the so-called Turing machine, can only be used for solving problems by a formal approach.

Penrose explains:

> The point of view that one can dispense with the meaning of mathematical statements, regarding them as nothing but strings of symbols in some formal mathematical system, is the mathematical standpoint of formalism. Some people like this idea, whereby mathematics becomes a kind of 'meaningless game'. It is not an idea that appeals to me, however. It is indeed 'meaning'—not blind algorithmic computation—that gives mathematics its substance.[10]

In drawing a conclusion from his meta-mathematical observations, Penrose leaves mathematics and enters the field of anthropology. And his conclusion is that the human mind is able to carry out certain activities that artificial devices, including the Turing machine, can not. It is not the question whether the human mind is superior in all aspects to the Turing machine, for example in computing speed even present-day computers exceed by far human capacity. The remarkable is that man can do something, for example grasping the meaning of a mathematical truth, a machine never can do. Penrose writes:

> Mathematical truth is not something that we ascertain merely by use of an algorithm. I believe, also, that our consciousness is a crucial ingredient in our comprehension of mathematical truth. We must 'see' the truth of a mathematical argument to be convinced of its validity. This 'seeing' is the very essence of consciousness. It must be present whenever we directly perceive mathematical truth. When we convince ourselves of the validity of Gödel's theorem we not only 'see' it, but by so doing we reveal the very non-algorithmic nature of the 'seeing' process itself.[11]

The significance of this 'seeing' a truth becomes even clearer if one uses the equivalent Latin verb *intueri* or the derivative 'intuition'. The latter points to a special intellectual activity that nearly instantaneous in a crea-

---

[10] Penrose 1989, 105.

[11] Penrose 1989, 418.

tive way is able to understand or conceive a solution to a problem. The foundation of this peculiar intellectual capacity is, according to Penrose, related to consciousness. One should, however have in mind that investigating this foundation one is not working any more in the realm of mathematics or even science in general, but more within anthropology or philosophy of man and metaphysics.

What about artificial intelligence? In the light of Penrose's conclusions one could speak of a misleading term. A person who is able to memorize very well or to solve certain formal problems in a rapid way, but who is missing insight and understanding is not considered to be intelligent. And the artificial devices like the Turing machine exclusively based on physical laws, can do it perhaps more rapidly, but with the same fundamental lack of insight. In stead of AI one should speak of ADP: artificial data processors.

## *5. Philosophical evaluation*

As already stated in the introduction, the ultimate goal of the philosopher is to find an overall consistent view where all empirical knowledge could be integrated. With the work of Gödel, Turing and Penrose one is confronted with two different types of intellectual activity in mathematics, which both can be carried out by humans. The astonishing thing is not the distinction between a formalized, logic approach on the one side and intuition, mathematical insight and meaning on the other. Being surely relevant for the philosophy of mathematics it does not have strong implications for the overall view on reality. Philosophically challenging, however, is the claim that principally only one of these approaches can be carried out by devices bound exclusively to the laws of physical reality. Turing machines, namely, are not able to grasp issues related to mathematical meaning, and these machines are the best physical laws can offer

Accepting the evidence provided by mathematics the philosopher should allow for two aspects of reality in his metaphysics: one exclusively bound to physical laws and the other not. A plain, ordinary materialistic view considering reality restricted to things that are exclusively subjected to physical laws is not able to incorporate the distinction provided by mathematics. The well-known quote by Karl Marx, *Der Mensch ist, was er ißt,*

(The human is that what he eats) is an example of a metaphysical statement that refutes this distinction: a human being is like his food, i.e. something exclusively bound to physical laws of nature.

Accepting the distinction suggested by mathematics the philosopher could provide more information about the ontological base of humans that allows them carrying out intellectual activities that Turing machines and other objects subjected exclusively to physical laws are not able to do. The principle of causality expressed in the classical statement *agere sequitur esse* (acting is according to the way of being) indicates that if an activity is observed that exceeds the potential physical laws can offer, there should be a subject exceeding the potential physical laws can offer. Already some of the Greek philosophers like Plato and Aristotle made a distinction between a material and spiritual reality. Especially Aristotle and later Aquinas studied the intellectual activity of human beings and found, interesting enough, a similar distinction in two principal approaches: strict logical argumentation with syllogism, a kind of formalized, deductive or inductive mental activity and intuition or in Greek *theoria* that accounts for insight and creativity.

It is interesting to note that in the metaphysics of Aquinas an intellectual activity is possible in pure spiritual beings and in man, who is composed of a spiritual and material component. According to Aquinas it is the material component in man that is responsible for the formal, logic intellectual activity with the aid of syllogism, whereas pure spiritual intellects know only by intuition. This is a remarkable parallelism with the findings in mathematics. A Turing machine, being a pure material device, can do the formalized processing, whereas only the human mind with his spiritual dimension according to Aquinas is able to carry out the intuitive intellectual activity needed to grasp mathematical meaning.

Aristotle like Aquinas did not know the Gödel theorem but arrived by a different analysis at the same conclusions as Penrose. This is, of course, no proof of the validity of the metaphysics of Aristotle and Aquinas, but an important consistency check with experience. There are, of course many other checkpoints, but after Gödel, Turing and Penrose this specific one provided by mathematics should be taken seriously, leading in some cases to the abandonment of certain metaphysical views, as, for example, provided by plain materialism.

## *Bibliography*

CACAE, F.

1997 "Meaning, Reality and Algorithms: Implications of theTuring Theorem" in Driessen, A. & Suarez, A. (eds.): *Mathematical Undecidability, Quantum Nonlocality and the Question of the Existence of God*. Dordrecht: Kluwer Academic Publishers.

GÖDEL, K.

1931 "Über formal unentscheidbare Sätze der Principia Mathematica und verwandter Systeme I" in *Monatshefte für Mathematic und Physik* 38.

HILBERT, D.

1935 *Gesammelte Abhandlungen*, III. Band. Berlin: Springer Verlag.

1900 *Mathematical Problems* (tenth problem). Paris: 2nd International Congress of Mathematics.

PENROSE, R.

1989 *The Emperors New Mind*. Oxford: Oxford University Press.

SCHINS, J. M.

1997 "Mathematics: A Pointer to an Independent Reality" in Driessen, A. & Suarez, A. (eds.): *Mathematical Undecidability, Quantum Nonlocality and the Question of the Existence of God*. Dordrecht: Kluwer Academic Publishers.

2000 *Hoeveel geest kan de wetenschap verdragen?* Kampen: Agora.

TURING, A.

1936-37 "On Computable Numbers with an Application to the Entscheidungsproblem" in *Proceedings of the London Mathematical Society*, 42-43.

# Informational Universe
# A Praxeo-Onto-Epistemological Approach

Wolfgang Hofkirchner, Christian Fuchs & Bert Klauninger

The spread of information and communication technologies in the course of the development of societies can be said to have triggered an "informational turn" in science and humanities. The concept of information and related concepts are widely used on both science and humanities. Thus the informational turn sets the stage for overcoming the divide between the two cultures and restituting metaphysical concepts brought forth in Antiquity, in particular, by Aristotle, albeit on a higher level. The universe can be conceived of as something that is self-organising. Introducing self-organisation is giving a new answer to the problem of how to relate inert matter on the one hand and living matter and mind on the other. This article will reveal epistemological, ontological and praxeological implications of this view of the informational universe.

## *1. Ways of Thinking in Cosmologies*

Human approaches to the universe may be distinguished according to the ways of thinking they reveal. Speaking in abstract terms, ways of thinking concern the consideration of how to relate identity and difference.

There are, in terms of ideal types, several ways conceivable:

- either the different sides are considered identical, that is, either side of the difference may be regarded as the base of identity; hence one way of thinking establishes identity by eliminating the difference at the cost of the differentiated side, it reduces the differentiated side to the undifferentiated one what is known as reductionism; another way of thinking establishes identity by eliminating the difference at the cost of the undifferentiated side, it takes the differentiated side as its point of departure and projects (extrapolates) from there to the undifferentiated one and may be called projectionism;

- or the different sides are considered different, that is, all relationships between them are abandoned; this way of thinking establishes the difference by eliminating identity; it dissociates both sides of the difference and treats them as disjunctive; this is what dualism (pluralism) is about;
- or they are considered to be both identical and different so as to establish identity in line with the difference; this way of thinking integrates both sides of the difference and differentiates the identical at the same time; it is a way of thinking that is based upon integration and differentiation and may be termed dialectic as it mediates opposites.

In the case of approaching the universe from a human perspective ways of thinking gain a more concrete meaning: the universe plays the role of that which is identical and the human and the nonhuman play the role of the two sides of the difference; the question that arises is how they make up the universe. This goes hand in hand with the question of how humanities and social science, on the one hand, may or may not be linked to the so-called exact science, on the other, and may or may not contribute to the unity of science.[1]

Reductionist approaches reduce the human side of the difference to the nonhuman side and set up the unity of the universe as a natural one. Hence their name: naturalism.

Projectionistic approaches use the human side as their starting point and project properties of the human side onto the nonhuman one, thereby unifying the universe on the basis of anthropomorphisms. These approaches are anthropomorphistic.

Dualistic approaches dissociate the human and the nonhuman side and let the universe fall apart. The anthropic and the natural form distinct worlds due to the so-called two-cultures-thinking.

Dialectical approaches integrate the human with the nonhuman side while they differentiate the totality of the universe. The anthropic and the natural are entangled, interwoven, permeated in a complicated way that is described by dialectics.

Though naturalistic approaches suffer from the underlying naturalistic

---

[1] HOFKIRCHNER 1999.

fallacy – that pretends to deduce explananda describing human phenomena from premises describing natural laws –, they yet form the materialistic mainstream of today's cosmology. Weak and strong formulations of the so-called anthropic principle which, in turn, are trapped in the anthropomorphistic fallacy – that postulates as explanans for natural phenomena premises that include concepts that apply to the human sphere – form the idealistic counterpart of materialistic cosmology. Another idealism that would restrict natural laws to nature and human regularities to humans would not qualify for being regarded as cosmology.

The argument we want to bring forth is that the paradigm shift from the so-called classical, mechanical, Newtonian view of the world to the world view of the sciences of complexity, of theories of dynamic, open, nonlinear, adaptive and evolutionary systems – a secular change that has been underway since the sixties of the last century and has been penetrating discipline by discipline – seems to undermine reductionistic or projectionistic approaches that by way of short cuts transfer concepts from nature to humans or from humans to nature as well as views that claim an unbridgeable divide. This shift rather seems to underpin a cosmological view that mediates between nature and humans by undertaking generalisations which start from the particular at either side and are directed towards the universal comprising both nature and humans and by producing reverse trains of thought that arrive at the specialised as well. This is so because findings in the field of complex systems and concepts of self-organisation do support a dialectical way of thinking. We want to show how these findings and concepts are ready for being generalised in that way.

## *2. The View of the Self-Organizing World*

We live in an age of global problems that concern the survival of humanity. As it is in the nature of these challenges to be complex, they have to be approached in a similarly complex fashion. A paradigm shift as far-reaching as never seen before is under way. What is known as sciences of complexity, self-organisation theories, evolutionary systems theories, is an element, if not the core, of this overall shift.

According to this thinking, all science serves to support efforts to master the global challenges. According to it, more and more researchers discover evolutionary systems no matter which real-world object they may be investigating, for the provision of specialised knowledge about the functioning of different self-organising systems is essential to influence them in such a way as to trigger the most promising development paths. Finally, according to it, diverse methodological approaches are less and less viewed as impediments that endanger the unity of science; rather, they are increasingly regarded as useful means towards the same end and as an enrichment of science as long as the common basis of the different methods is not violated.

Before we present results of the sciences of complexity and make an attempt of generalisation we want to introduce the categories along which we will classify them. These categories stem from philosophy.

### 2.1 Praxeological, ontological and epistemological assumptions revisited

Philosophy is the most general reflection of humans and their position in the world. There are, at least, three fundamental questions that have been constituting philosophy since ever, though they have been put in different ways. One question is about values, norms, imperatives, guidelines for acting. Another question is about the world as it is, its properties, be it with or without us. A third question is about our ability to produce knowledge. The first question makes up the domain of ethics, aesthetics and axiology, the second makes up the domain of ontology, and the third the domain of epistemology including the methodology of inquiry.

These three domains may be tackled either as separate fields of philosophy, or as networked and even nested. Historically speaking, awareness of interdependencies of the fields has been growing.

After times of prevailing realistic stances in ontology, varieties of radical constructivism arising in the second half of the last century have put emphasis on the interdependence of epistemological and ontological questions (though some of them tended to end up in mere solipsism). According to them the world as it is is difficult to approach since our assumptions of how the world is turn out to rely on specific methods of human cognition. A moderate stance trying to reconcile realism with constructivism is onto-epistemology as coined by Hans Jörg Sandkühler[2] and shaped by Rainer Zimmermann.[3] We tie up to this strand and want to complement the interrela-

---

[2] Sandkühler 1990, 1991, 34-37, 353-369.

[3] E.g. Zimmermann 2002, 147-167.

tionship of ontology and epistemology by the relation to ethics, aesthetics and axiology all of which we propose to include in so-called praxeology.

Praxeology is meant as the philosophical theory of praxis. Praxis refers to society and its human actors. It can be defined as the process of co-action (of human co-actors) upon reality which, in turn, can be defined as the field of interaction (of human actors with the environment) that is mediated by method which, in turn, can be defined as the way of action (of those human actors). In terms of subjects and objects, praxis is the totality of the human subject-object-dialectic, reality is what is, so to say, objecting to becoming subject to humans, and method is the subjective way of making objects subject to humans. It becomes clear from that order of definitions that praxis builds upon reality and that reality builds upon method, be it material or ideational. Hence, if ontology is meant as the philosophical theory of reality and epistemology as the philosophical theory of the method of inquiry, praxeology is based upon ontology and ontology on epistemology. This relationship of encapsulation of domains of philosophical disciplines, and of the philosophical disciplines themselves, does not model a one-way process of causal influence or linear inferences. Both bottom-up as well as top-down processes have to be recognised. There is relative autonomy of each of the domains (praxis may shape reality but reality gives the scope of possible practices, reality may shape method but method gives the scope of possible realities) and of each of the disciplines (praxeology does not fully determine ontology and ontology does not fully determine epistemology, and vice versa).

The rationale for defining subject matters and respective theories in such a concatenated way is that we try to give an appropriate sketch of the following relations: certain interests (that reflect certain practices) define the sphere of intervention (that is made up of objects in which subjects are interested and is characterised by a boundary beyond which there are no real objects as long as there is no subject interested in them) and certain spheres of intervention (that reflect certain realities) define the scope of instruments (that is made up of means which are useful for intervention and is characterised by a boundary beyond which there are no real means as long as they do not fit the object); and, in turn, certain instruments (that reflect certain methods) can help construct a certain sphere of intervention only by which different realities are excluded and certain spheres of intervention can meet a certain bunch of interests only by which different interests are excluded.

Thus, taking explicitly a human stance, we can reformulate the fundamental questions of philosophy by starting with the praxeological question and subsequently introducing the ontological question and the epistemological question, each one being the presupposition for the question before:

1. How should the world be like?
2. How can humans make the world how it should be like?
3. How can humans understand how they can make the world how it should be like?

Having guidelines for action presupposes having ideas about where human actors start from and having ideas about where human actors start from presupposes having tools to recognise the starting point. If humans want to succeed in changing the world they need to know about the relations that allow to achieve the goals they have set. And in order to gain this knowledge they are obliged to apply all means that seem worthwhile. So our praxeo-onto-epistemological standpoint is indeed one in which praxeology does matter: ontic propositions bear the stamp of practical instructions and pass this stamp over to epistemic methods.

It becomes clear that by defining praxis as the subject matter of praxeology we denote by this term the philosophical theory of human actions in regard not only to their efficiency, effectiveness and efficacy but moral value and beauty as well.[4]

Our way of encapsulating praxeology, ontology and epistemology may be regarded as generalisation of what action theory is pointing at when establishing a relationship between ends, ways and means. Ends refer to practical interests, ways refer to ontic interventions, and means refer to epistemic instruments.

Furthermore, the way we conceive of the nestedness of praxeology, ontology, and epistemology resembles the way semiotics may be structured. Semiotics is about signs and may be divided into pragmatics, semantics and syntactics as subdisciplines. According to the perspective advocated here semantics is a subset of pragmatics and syntactics is a subset of semantics. Pragmatics deals with the use of signs, semantics is about the signs' relation to their referents and syntactics talks about signs in themselves and how their components fit together. If a sign is something that is produced and/or used by someone with the purpose to refer to something else in or-

[4] See different views in MISES 1999, KOTARBINSKI 1965, BUNGE 1999, COLLEN 2003.

der to express the meaning of it to someone else, then it makes sense to consider the pragmatic aspect as uppermost level, the semantic aspect as intermediate level and the syntactic aspect as the bottom level since the usage defines the reference and the reference defines the production of signs while the production gives rise to a certain scope of possible references which, in turn, gives rise to a certain scope of possible usages. If we look upon sign production as method, if we look upon the sign referents as reality and if we look upon the sign use as praxis, this analogy between praxeology-ontology-epistemology and pragmatics-semantics-syntactics seems not casual.

There may also be historical evidence for the subsequent differentiation of the philosophical disciplines of praxeology, ontology and epistemology. A first stage of development may be the existence of mythology which in ancient societies was a reflex of human praxis as an undifferentiated whole. Another stage of development may be characterised by the advent of the philosophy of nature produced in greek civilisation and contemplating the object of practices. A third stage of development, at last, may be introduced by specialised philosophy of science considerations which initiated the self-reflection of human inquiry.

## 2.2. Universals of the sciences of complexity

Praxeology, ontology and epistemology form the categories along which we can try to classify the results of the sciences of complexity. In our terms, sciences of complexity have one central focus: self-organisation. Self-organisation is the most distinctive character of systems that have the capacity to contribute to evolution: it is the process and result of the spontaneous build-up of order. To recognise the self-organising capability of matter, of nature, of real-world systems has implications for praxeology, ontology and epistemology.[5]

[5] See Hofkirchner 2004.

### 2.2.1. Praxeological results and generalisations: Evolutionary Systems Design Principles and Deliberate Activism

In a systems theory perspective, dealing with complexity is located inbetween two poles – the pole of algorithmisation and the pole of nonprogrammability. Algorithmisation means the process of developing clear-cut and unambiguous instructions that can be carried out by the help of computers as universal machines. Up to now the bulk of technologies has been designed according to such a rule, but disasters increasingly render this view inappropriate. Nonprogrammability refers to just this experience that systems in nature do not obey that rule and escape simulation and intervention and cannot be reproduced.

Evolutionary systems design principles which can be abstracted from the wide range of findings in system research avoid to get deadlocked at one of the poles. They encourage to make use of the systems' dynamic and stress the point that knowing about nonlinearity and sensitivity may help to choose those inputs that trigger developments in the overall self-organisation process of the system that are favourable to those who make the inputs. System processes may be facilitated or may be dampened. Also it is important to influence the general set-up of the system only and abandon instructions down to every detail so that relative autonomy is granted to the subsystems. All in all, this kind of systemic "intervention" can be labeled "decentralised context steering".[6]

In a philosophical perspective, decentralised context steering finds its general pendant in a kind of piecemeal engineering Karl Raimund Popper[7] introduced to the terminology of social sciences or in an attitude Francis Bacon[8] advocated in his saying that changing the world requires observing its laws, if we – today – include the law of unintended consequences.[9]

This kind of activism is not a practicism that guides action according to the maxim that all that is feasible shall be realised thereby assuming that it is desired too. Nor is this kind of activism a utopian or romantic wishful thinking that holds that what is desired is feasible too. Both practicism and wishful thinking believe in total controllability and result in expensive brute-force

---

[6] Jessop 1997, Teubner and Willke 1980, Willke 1995.

[7] Popper 1962.

[8] Bacon 1990.

[9] Merton 1936, 1957.

interventions. Nor is this kind of activism an inactivism that believes in total uncontrollability, condemns any kind of intervention and fails to reconcile the feasible and the wishful. It is deliberate activism that takes responsibility for producing the unity of the feasible and the wishful. And it does so by working out the ascendence from the here-and-now to the not-yet in the sense of Ernst Bloch[10] and the ascendence from the less good to the better.

### 2.2.2. Ontological results and generalisations: Evolutionary Systems Stage Model Principles and Less-than-strict-Determinism

Mechanism and spontaneity are the two extremes taking complexity as basis can be trapped in. Mechanism refers to one-way cause-effect-relationships underlying the systems' processes and structures while spontaneity denies this and ascribes irreversibility to the processes and irreducibility to the structures. The computer metaphor insinuates the whole world is working like a computer while, on the other hand, only less-than-middle-range narratives are said to be the best to be achieved.

The stage model of evolutionary systems, however, is based upon the principle of emergentism and the principle of asymmetrism. Emergence takes place in metasystem transitions in which by the interaction of proto-elements systems are produced[11] , evolution is characterised by punctuated equilibria.[12] Asymmetry describes the architecture in which higher levels are based upon lower ones but nevertheless exert a downward causation in turn, supersystem hierarchies encapsulate subsystems in the way Arthur Koestler talked about holons.[13]

If we try to generalise these system theoretical findings and raise them onto a philosophical level, we are confronted with the determinism-indeterminism divide. There is no determinacy without indeterminacy and no indeterminacy without determinacy. The French philosopher Edgar Morin aptly coined the term "chaosmos" to indicate that the world is cosmos and chaos at the same time.[14] When Popper talked about propensities in one of his last publications, he had in mind the very same idea of a universe that

---

[10] BLOCH 1985.

[11] JOSLYN, HEYLIGHEN, TURCHIN 1997.

[12] ELDREDGE and GOULD 1972, GOULD and ELDREDGE 1977.

[13] KOESTLER 1972.

[14] MORIN 1992, 53–64.

is inclined to exhibit some properties[15] rather than governed by strict natural laws. The American pragmatist Charles Sanders Peirce said that evolution takes habits.[16] One of us called this idea of weak determination elsewhere less-than-strict-determinism.[17]

Less-than-strict-determinism is not a preformationism according to which evolution is only unfolding of something already existing and not a merism according to which wholes can be reduced to their parts. Both preformationism and merism reduce chance to necessity. Less-than-strict-determinism is not teleologism or holism which project chance in the form of a not existing goal or a contingent whole onto necessity. And it is not a dichotomism of necessity and chance. It tries to work out the unity of necessity and chance by the ascendence from the old to the new and from the parts to the whole.

#### 2.2.3. Epistemological results and generalisations: Evolutionary Systems Methodology Principles and Reflexive Rationalism

Methods of thinking in complexity are exposed to the tension between formalisation and the nonformal. Formalisation is the activity of symbolising a real-world process in order to be able to operate on the symbols while taking into consideration rules of combination of the symbols only and neglecting their referents. Having carried out the operation you can refer to the referent again. Formal logic and mathematics are instances of such languages. They can work if and because the real-world referents are in reality linked in the strict deterministic way formal logic and mathematics link symbols. Systems modelling always confronts with situations that are not that way formalisable and hence unpredictable. Computer science knows about the problem of the nonformal.[18] Hence the call for qualitative methods as opposed to quantitative ones. Qualitative methods are said to be the only appropriate for the nonformalisable.

Emergence in phases of evolution, including the appearance of novel qualities in developments, and differences between system levels cannot be

15 POPPER 1997.

16 PEIRCE 2000.

17 HOFKIRCHNER 2001.

18 See FUCHS-KITTOWSKI 1992.

formalised in a way that there is a transformation that leads unequivocally from one to another. Evolutionary systems methodology therefore has to recognise the principles of formalisation gaps: in the case of phase transitions where a leap in quality exists between the state of the system at one point of time (t) and the following state (at time t+1) and in the case of level shifts where a leap in quality exists between one layer of the system and the adjacent layer up or down (micro- and macrolevel). It holds it is impossible to find an operation in the mind that accomplishes the leap from one quality to another in an unambiguous and compelling way.

However, in a general philosophical perspective, there is a terminology that allows to mediate different qualities. With the help of this terminology the one can be described as the condition for the other and the other, then, can be described as the conditioned. This is the way out of the deductivism-irrationalism chasm. Deductivism favours complete deducibility while irrationalism is willing to accept any nondeducible statement. Within deductivism you may distinguish analytical thinking with a naturalist bias and synthetical thinking with an anthropomorphistic bias. The first reduces the historical to the logical (saying laws in nature are nothing but intelligible necessities), the second projects the historical onto the logical (saying laws are nothing but contingent). Irrationalism underlines the opposition of the logical and the historical.

What we would like to term reflexive rationalism abstains from providing causal explanations and predictions which are conclusions drawn from premises as deductive rationalism would do and it refrains from disseminating narratives that deny inferences as irrationalism would do. Reflexive rationalism establishes the unity of the logical and the historical by announcing the principle of the search for the approximate necessary, but not in all cases sufficient, condition. The approximate necessary condition is the representation of the real prerequisite for the quality to become, be it in phases or in levels. The ascendence from the potential to the actual as well as from the abstract to the concrete is carried out by jumps from the condition to the conditioned.

## *3. The Rise of the Informational Universe*

The philosophical generalisations of the results of the sciences of complexity are useful to approach an answer to the question of which place information has in the universe and which role creativity plays.[19]

There is evidence for the fact that the universe humans inhabit has been undergoing a process of evolution at least since 13.7 billion years. Given the insights of the philosophical underpinning of self-organisation theory, it is hard to fancy a beginning like the Big Bang. For "nothing will come of nothing". As far as we know, each stage in any evolutionary process is prepared by a preceding stage in that the preceding one builds the foundation for the next irrespective of what the next stage will be like. Hence the stage before the first stage human science acquires knowledge of is not "nothing" but "something else" – an object of human imagination and inquiry. This is an argument for the eternity of the overall evolution put forward by prominent materialist thinkers as Friedrich Engels.[20]

Evolution apparently reaches ever-higher complexity which, however, is recovered by ever-new simplicities. The points of bifurcation that are passed from stage to stage may be deciphered as symmetry breaks. So self-organisation theory serves well in interpreting the how of the evolution of the universe.[21]

## 3.1. Information and self-organisation

Information seems to be as fundamental a building block of the universe as matter and energy is. So far, however, there is no common agreed-upon concept of information.[22] There are objectivistic approaches, originating from the so-called "hard" sciences, that consider information a substance because of its ability to be received and processed in cognitive contexts, to be transmitted in communicative contexts, and to be stored and retrieved and distributed in co-operative contexts. On the other hand, there are subjectivistic approaches, originating from the so-called "soft" sciences, that believe information to be a mental construct produced by human actors internally, or interactively, or externally according to the context.

---

[19] Esoteric answers range from an early book of YOUNG 1974 to a recent book of GÖRNITZ and GÖRNITZ 2002.

[20] WOODS and GRANT 2002, 177-222, see FUCHS 2003.

[21] See e.g. EBELING and FEISTEL 1994, KANITSCHEIDER 2002, 467-474, SMOLIN 1997, SMOLIN 2003, 201-205, LAYZER 1990.

[22] See CAPURRO and HJØRLAND 2003.

An integrative information concept cannot be satisfactorily formed on the basis of such one-sided views. What we call the Unified Theory of Information perspective has to consider both the objective and subjective aspects of information which have to be related to each other in a proper way. The objectivist outlook is right in stating that information is a phenomenon out there (an object) and not merely a human imagination. The subjectivist outlook is right insofar it states that information occurs only if there is a freedom of choice for the thing to which generating and disposing of information is attributed (a subject). However, regarding the objectivist outlook, we have to limit the scope of objects with which information is said to be found to those objects only that take the role of subjects, and regarding the subjectivist outlook, we have to enlarge the sphere of subjects from that of humans only and include non-human ones, too.

According to this unifying perspective, information is a relation of self-organising systems to their environment (cognitive context), to their co-systems (communicative context), or to the supersystem they give rise to (co-operative context). In terms of evolutionary systems theory, information is generated iff self-organising systems relate some external perturbation to the spontaneous build-up of order they execute when exposed to this perturbation. To translate into terms of triadic semiotics, by doing so, the self-organising systems assign a signification to the order and make it a sign which stands for the so signified perturbation. Thus the process of self-organisation coincides with the process of information-generation (sign-production) and so do their respective results. The concepts of self-organisation and information turn out to be co-extensive.

## 3.2. The rise of subjectivity

Speaking of subjectivity means returning to the terms subject and object. According to our praxeo-onto-epistemological point of view, the difference between subject and object may be seen in that a subject is capable of determining itself while an object is not. An object is something that is determined by something that is not itself. Being a subject supersedes being merely an object. While an object has no possibility of acting in ways different from merely reacting to external determinants, a subject is capable of responding in its own, unequivocal way, that is, it can make use of degrees of freedom, of freedom of choice, of choice between options all of which it disposes of and thus makes the internal determine. It may object to external determinants while objects do not.

So, if an object is something that is subject to mere determination by something else and if a subject is something that objects to mere determination by something else, then information-generating, self-organising systems display a certain subjectivity, for the generation of information is tantamount with drawing a self-made distinction by the irreproducible, irreversible, irreducible, unpredictable build-up of order during the process of self-organisation.

The minimal unit of subjectivity is a something that is provided with a minimal quantum of degrees of freedom to act. This something is the most rudimentary and most primitive subject.

The making of something subject to oneself which makes oneself a subject undergoes a process of unfolding so as to let us distinguish between different types of subjects according to the degree of subjectivity they manifest. It is different types of systems that show different degrees of subjectivity. The more complex a system, the more subjectivity it displays. This holds for the ascendence from physico-chemical self-organising (dissipative) through biotic (autopoietic) to social (re-creative) systems and for the progress of society as well. In the course of evolution, there is a drift from drivenness towards end-directedness and from materiality towards formative power.

Aristotle knew four types of causes: the effective (causa efficiens), the final (causa finalis), the material (causa materialis) and the formal (causa formalis) one. In a strive for scientificity that avoided resorting to the supernatural, post-medieval science abandoned the latter three of them. However, in the perspective of evolution and systems it is worth reconsidering all four types of causes without need to resort to the supernatural. You can arrange them on two axes. One axis shows the diachronic dimension of systemic evolution and goes from drivenness to end-directedness, another shows the synchronic dimension of evolutionary systems and goes from materiality to formative power.[23] You can arrange the effective and final cause on the first axis and the material and formal cause on the second one.

---

[23] BRUNNER and KLAUNINGER 2003.

Effective cause connotates a driving force in the process while final cause connotates rather to pull than to push. But finality does mean influence "from the future" as little as effectivity means exerting pressure "from the past". Each process paves the way for the future by its own history. It brings into existence a certain range of possibilities and a complementary range of impossibilities. Those possibilities do exist in the present and one of them will be selected and realised and then open up another range of possibilities. Compared to the range of impossibilities, the process converges to one end after the other through a series of concatenated ranges of possibilities.

And material cause connotates the substantial base in the structure while formal cause connotates the shaping of it. And formality does mean influence "by consciousness" as little as materiality means exerting pressure "by matter". Each structure bears the stamp of how its constituents compose it. The constituents produce what they constitute by producing constraints as well as enablers which represent the form.

The evolutionary systems stage model assumes that it is already the most primitive type of self-organisation that not only exhibits effective and material causation but also to some extent final and formal causation. Dissipation realises some end (as to the energy throughput) as well as some form (by which the energy throughput is served) though not consciously. Dissipative systems may hence be termed "teleomatic"[24] with regard to the primitive type of end-directedness they exhibit (they automatically seem to realise an end) and "self-referential" with regard to the primitive type of formative power they exhibit (they refer to themselves in that there is a feedback established from the macro- to the microlevel). Autopoiesis goes beyond mere dissipation. It does not only realise some end but inheres it, the end is implicit (survival for any living system). Hence autopoietic systems may be termed "teleonomic". Autopoiesis furthermore does not only realise some form but reproduces it time and again (the organisation of any living system). Hence autopoietic systems may be termed "self-maintaining". And last not least does re-creation go beyond mere autopoiesis. It not only implies an end but makes it explicit (goals are set) and it not only reproduces the form but creates it (the mode of production of any social system is artificial). Hence re-creative systems may be termed "teleologic" in the full sense of the word and "self-inventive".

[24] See MAYR 1988.

Thus we can discern degrees of subjectivity. We can depict simple dissipative systems as proto-subjects and simple autopoietic systems as para-subjects while reserving the property of being a subject in the full sense of the word for re-creative systems only. But there is a continuum in the evolution of subjectivity. In this sense the word is true that in the course of evolution by means of ever-higher developed natural systems like social systems nature more and more comes to reflect itself.

And societal evolution is no exception from the rule. With information age societal evolution can be said to approach a point of bifurcation that allows evolution of consciousness to shift to conscious evolution. Jonathan Salk put it already in 1983:

> The most meaningful activity in which a human being can be engaged is one that is directly related to human evolution. This is true, because human beings now play an active and critical role not only in the process of their own evolution but in the survival and evolution of all things. Awareness of this places upon human beings a responsibility for their participation in and contribution to the process of evolution.[25]

Bela H. Banathy adds:

> If we accept this responsibility and engage creatively in the work of evolution we shall take part in a crucial and a first ever event in the seven million years of our evolutionary saga: We shall be the designers of our future, we shall become the guides of our own evolution and the evolution of life on earth and possibly beyond.[26]

These ideas resemble the ideas of the noosphere to come coined by Pierre Teilhard de Chardin[27] and Vladimir I. Vernadsky[28] in the time between world war I and II.

This shift is, so to say, the progressive, upper branch of the great bifurcation of human history and the history of the cosmos as well, the regressive, lower branch of which might end up in decline and decay if humankind is not able to make up with the gap between the technological and the social evolution.

---

[25] Salk 1983, 112..
[26] Banathy 2000, 203.
[27] de Chardin 1975.
[28] See Hofkirchner 1997.

> While our recently emerged communication capabilities created the potential and the conditions for global human community, our consciousness is still locked within ethnocentric, racial, and national boundaries. ... Furthermore, the technological revolution, while giving us an earlier unimagined power, has accelerated to the point where we have lost control over it.[29]

> We have simply failed to match the advancement of our technological intelligence with an advancement in socio-cultural intelligence, and advancement in human quality and wisdom.[30]

The great bifurcation takes the form of the choice between the breakthrough to a higher level of organisation of humankind and its breakdown. Thus we can conclude: praxeologically, in order to open up space for anthropo(socio)genesis as part of the cosmogenesis to turn into noogenesis and to avoid exterminism, humanity requires strategies to master the great bifurcation which is, ontologically, due to the vulnerability of the technosphere, the limited carrying capacity of the ecosphere, and the economic, political, and cultural exclusiveness of the sociosphere and has to be researched by the transversal efforts of the sciences of complexity.

## *Acknowledgements*

This article was written in the context of the research project „Human Strategies in Complexity: Philosophical Foundations for a Theory of Evolutionary Systems" funded by INTAS (The International Association for the Promotion of Cooperation with Scientists from the New Independent States of the former Soviet Union) and the Austrian Federal Ministry of Education, Science and Culture.)

## *Bibliography*

ALSLEBEN, K. & WEHRSTEDT, W. (Eds.)
1966 *Praxeologie*. Quickborn: Schnelle.

BACON, F.
1990 *Weisheit der Alten*. Frankfurt: Fischer.

[29] BANATHY 2000, 193.
[30] BANATHY 1996, 315.

BANATHY, B. H.

1996 *Designing Social Systems in a Changing World.* New York: Plenum Press.

2000 *Guided Evolution of Society, A Systems View.* New York: Kluwer.

BLOCH, E.

1985 Das Prinzip Hoffnung. Frankfurt am Main: Suhrkamp.

BRUNNER, K. & KLAUNINGER, B.

2003 "An Integrative Image of Causality and Emergence" in: Arshinov, V. & Fuchs, C. (eds.) *Emergence, Causality, Self-Organisation.* Moscow: NIA-Priroda.

BUNGE, M.

1999 "Ethics and Praxiology as Technologies" in: *Techné* Vol. 4, No. 4 (http://scholar.lib.vt.edu/ejournals/SPT/v4n4/bunge.html#kotarbinski).

CAPURRO, R. & HJØRLAND, B.

2003 "The Concept of Information" in: Cronin, B. (ed.) *Annual Review of Information Science and Technology.* New Jersey: Medford.

DE CHARDIN, P. TEILHARD

1975 *The Phenomenon of Man.* New York: Harper & Row.

COLLEN, A.

2003 *Systemic Change Through Praxis and Inquiry.* New Brunswick, London: Transaction Publishers.

EBELING, W. & FEISTEL, R.

1994 *Chaos und Kosmos, Prinzipien der Evolution.* Heidelberg: Spektrum.

ELDREDGE, N. & GOULD, S. J.

1972 „Punctuated equilibria, An alternative to phyletic gradualism" in: Schopf, T. J. M. (ed.) *Models in Paleobiology.* San Francisco: Freeman, Cooper. 82-115.

FUCHS, C.

2003 "Dialectical Materialism and the Self-Organisation of Matter" in: *Seeking Wisdom,* Vol. 1, No. 1. 25-55.

FUCHS-KITTOWSKI, K.

1992 „Theorie der Informatik im Spannungsfeld zwischen formnalem Modell und informaler Welt" in: Coy, W. (ed.) *Sichtweisen der Informatik.* Braunschweig, Wiesbaden: Vieweg. 71-82.

GOULD, S. J. & ELDREDGE, N.

1977 „Punctuated equilibria, The tempo and mode of evolution reconsidered" in: *Paleobiology,* 3. 115-151.

GÖRNITZ, T. & GÖRNITZ, B.
2002 *Der kreative Kosmos, Geist und Materie aus Information.* Heidelberg, Spektrum.

HOFKIRCHNER, W.
1999 "Ways of Thinking and the Unification of Science" in: Allen, J. K. & Hall, M. L. W. & Wilby, J. (Eds.) *Proceedings of the 43rd Annual Conference of ISSS* (The International Society for the Systems Sciences). ISBN 09664183-2-8 (CD-ROM), 12.
2001 "The hidden ontology: Real-world evolutionary systems concept as key to information science" in: *Emergence*, Vol. 3, No. 3. 22-4.
2004 "A New Way of Thinking and a New World View: On the Philosophy of Self-Organisation I" in: *Cybernetics & Human Knowing*, Vol. 11, No. 1. 63–78.

HOFKIRCHNER, W. (ed.)
1997 *Vladimir I. Vernadskij: Der Mensch in der Biosphäre, Zur Naturgeschichte der Vernunft.* Wien: Peter Lang.

JESSOP, B.
1997 "The Governance of Complexity and the Complexity of Governance, Preliminary Remarks on Some Problems and Limits of Economic Guidance", in: Amin, A. & Hausner, J. (eds.) *Beyond Market and Hierarchy, Interactive Governance and Social Complexity.* Cheltenham: Edward Elgar.

JOSLYN, C. & HEYLIGHEN, F. & TURCHIN, V.
1997 *Metasystem Transition Theory.* http://pespmc1.vub.ac.be/MSTT.html.

KANITSCHEIDER, B.
2002 *Kosmologie.* Stuttgart: Reclam.

KOESTLER, A.
1972 *The Roots of Coincidence.* London: Hutchinson.

KOTARBINSKI, T.
1965 *Praxiology.* Oxford: Pergamon Press.

LAYZER, D.
1990 *Cosmogenesis, The Growth of Order in the Universe.* New York: Oxford University Press.

MAYR, E.
1988 *Toward a New Philosophy of Biology, Observations of an Evolutionist.* Cambridge: Harvard University Press.

MERTON, R. K.
1936 "The Unanticipated Consequences of Purposive Social Action" in: *American Sociological Review*, Vol. 1, Issue 6. 894-904.
1957 *Social theory and social structure.* Glencoe: Free Press.

MISES, L.V.
1996 *Human Action.* http://www.mises.org/humanactions.asp.

MORIN, E.
1992 *Method.* New York: Peter Lang.

PEIRCE, C. S.
2000 *Semiotische Schriften.* Bd. 1, 2, 3. Frankfurt: Suhrkamp.

POPPER, K. R.
1962 *The Open Society and its Enemies.* London: Routledge and Kegean.
1997 *A world of propensities.* Bristol: Thoemmes Press.

SALK, J.
1983 *Anatomy of Reality: Merging of Intuition and Reason.* New York: Columbia University Press.

SANDKÜHLER, H. J.
1990 „Onto-Epistemologie" in: *Europäische Enzyklopädie zu Philosophie und Wissenschaften.* Hamburg: Meiner.
1991 *Die Wirklichkeit des Wissens, Geschichtliche Einführung in die Epistemologie und Theorie der Erkenntnis.* Frankfurt: Suhrkamp.

SMOLIN, L.
1997 *The Life of the Cosmos.* New York: Oxford University Press.
2003 *Three Roads to Quantum Gravity, A New Understanding of Space, Time and the Universe.* London: Phoenix.

TEUBNER, G. & WILLKE, H.
1980 „Dezentrale Kontextsteuerung im Recht intermediärer Verbände" in: Voigt, R. (ed.) *Verrechtlichung.* Königstein: Athenäum.

WILLKE, H.
1995 *Systemtheorie III, Steuerungstheorie.* Stuttgart: Fischer.

WOODS, A. & GRANT, T.
2002 *Reason in Revolt, Dialectical Philosophy and Modern Science.* Vol. 1. New York: Algora Publishing.

YOUNG, A.
1974 *The Reflexive Universe.* Green Leas: Robert Briggs Associates.

ZIMMERMANN, R.
2002 *Kritik der interkulturellen Vernunft.* Paderborn: Mentis.

# Intuitive Universe: Immediacy in Human Knowledge

LASSI LARJO

## *1. Introduction*

The distinction between rational intuition (noesis) and intellectual discursive knowledge (dianoia) traces back to Ancient philosophy. The latter was considered to be dealing with the lowest parts of the intellectual cosmos, whereas direct intuition captured the most unshakeable, evident and immediate truths. In the Platonic tradition, intuition was the medium of eternal and universal ideas; Aristotle associated intuition with the apprehension of primary premises of scientific thought.[1] The division between intuitive and discursive knowledge has ever since been one of the major dualisms plaguing Western thought. Simply put, the question is whether human knowledge of reality is immediate and direct or mediate and indirect. Especially in the Platonic tradition is a strong current of the "metaphysics of vision" in which intuition, or knowledge through seeing or acquaintance, is essential. As the Latin etymology of the word "intuition" signifies[2], intuition is about *seeing* the intelligible correctly. However, the association of intuition with sensuous vision, with the "eye of the flesh", so to speak, is misleading, yet analogous. St. Augustine made this analogy quite clear:

> But we ought rather to believe that the nature of the intellectual mind is so formed as to see those things which, according to the disposition of the Creator, are subjoined to intelligible things in a natural order, in a sort of incorporeal light of its own kind, as the eye of the flesh sees the things that lie about in this corporeal light, of which light it is made to be receptive and to which it is adapted.[3]

---

[1] PLATO, *The Republic* 514-518; ARISTOTLE, *Posterior Analytics* 100b.

[2] From the Latin word *intueri*, "to look at", "to consider".

[3] ST. AUGUSTINE 1981, XII, xv, 24.

Intuitions are like visions of the soul seen with the mind's eye, which is something completely different from discursive reasoning. Discursive thinking requires inference and deduction – the elaboration of received information.

The role of intuition has been central in human approaches to the universe, but today its function is ambiguous. In order to understand what has happened to our appraisal of intuition, I believe it is crucial to investigate a few important developments in modern philosophical discussions concerning it. In this paper I will especially focus on 18th century thought in which the most substantial changes concerning the role of intuition took place.

## *2. Modern intuition before Kant*

Usually early modern philosophy valued intuition as highly as it was valued in Ancient thought. Descartes maintained that intuition is primarily a function of the intellect through which "everyone can mentally intuit that he exists, that he is thinking, that a triangle is bounded by just three sides and a sphere by a single surface, and the like."[4] This suggests that intuition has nothing to do with the outer senses.[5] However, Descartes could not avoid analogies that related intuition strongly to sensate vision. On the other hand, it is not at all clear what the difference is between intuition and deduction. Is the famous *cogito ergo sum*, for example, intuitive or discursive knowledge?[6]

Spinoza's thought strengthens the connection between intuition and mental seeing. According to his view, intuition is one of the four modes of perception that give us a full account of all kinds of knowledge. It is evident that when discussing intuition as perception Spinoza does not refer to "the eye of the flesh", which seems to be more appropriately associated

---

[4] DESCARTES 1985, 14.

[5] "By intuition, I do not mean the fluctuating testimony of the senses, or the deceptive judgement of the imagination as it botches things together, but the conception of a clear and attentive mind which is so easy and distinct that there can be no room for doubt about what we are understanding. Alternatively, and this comes to the same thing, intuition is the conception of a clear and attentive mind which proceeds solely from the light of reason." (DESCARTES 1985, 14).

[6] See e.g. HACKING 1986, 51-52 and the literature mentioned there.

with another mode, perception through "mere experience". Intuition is the most reliable source of perception, "the perception arising when a thing is perceived solely through its essence, or through the knowledge of its proximate cause."[7] Moreover, Spinoza seems to support the idea that the intuitive eye of the mind has the most pertinent access to reality.

Thus, it seems that early modern philosophers tended to associate intuition with intellectual vision. Truths apprehended intuitively were considered self-evident, and thus in a way common sensical.[8] It is also noteworthy that philosophers labelled "rationalists" usually held intuition as a pure and abstract apprehension. We might say, for instance, that Leibniz's Principle of Sufficient Reason is just as intuitive as it is formal and abstract. This way of understanding the role of intuition was changing, however. Some historians of philosophy suggest that a comprehensive conflict between intuition and reflection motivated most of the substantial philosophical problems of the late 18th century.[9] In the modern context it became more and more problematical to support any claims of intuitive knowledge in the traditional sense. I do not mean to suggest, that "modern" or some essential hallmark in modern thought in itself is deleterious to intuition. In fact, it seems plausible that the vast majority of early modern thinkers fought for intuition, against thought patterns that might endanger it. The mutual foe of the mainstream thinkers was what Richard H. Popkin has aptly called "the third force in 17th century thought", scepticism.[10] But in the end this defensive battle, almost by accident, turned out to be detrimental to intuition.

The role of British philosophy is remarkable in this process. It is suggested that the empiricists tried to extirpate the remains of the metaphysics of vision from all science and replace it with a more straightforward theory of experience. But in reference to intuition is this really the case? According to Locke, intuitive knowledge "is irresistible, and, like bright sunshine,

[7] SPINOZA 1985, P19. The other three modes are hearsay, mere experience and inadequate inferential perception.

[8] SPINOZA states in his *Ethics* (SPINOZA 1982), II:45, Scholium 2: "For example, in the case of the given numbers 1, 2, 3, everybody can see that the fourth proportional is 6, and all the more clearly because we infer in one single intuition the fourth number from the ratio we see the first number bears to the second."

[9] DI GIOVANNI 1992, 420.

[10] This is also the title of RICHARD H. POPKIN's excellent collection of essays discussing the role of scepticism (1992).

forces itself immediately to be perceived as soon as ever the mind turns its view that way; and leaves no room for hesitation, doubt, or examination".[11] Hume approximated Locke in his definitions. Unlike Locke, however, he identified intuitive knowledge exclusively with demonstrative knowledge, such as propositions of geometry, algebra and arithmetic.[12] His famous sceptical stance towards inductive reasoning was a fundamental denial of any intuitive connection between the terms of an inductive inference.[13] Criticism of alleged knowledge founded on causality is another form of this attack against reason and intuition. According to Hume reason is merely "the slave of passions". The epistemological trinity of reality, reason and intuition is, if not utterly false, unattainable from the human perspective.

The apodictic certainty attached to all events perceived in the cosmos seemed to reduce under closer examination to mere habit or custom. Such a conclusion was something that practically all philosophers tried to evade by all possible means. The question of a secure, intuitive foundation of human knowledge had to be saved. For Descartes, the *terra firma* of all reliable judgements was to be found from the undisputable – and, yes, according to some interpretations, intuitive – existence of the thinking self. But perhaps more typically the secure foundation of exhaustive philosophical systems was sought from mathematical exposition. In the end, it was the unquestionable mathematical form of reasoning that secured the famous *cogito* argument. Hence, it was typical to call the proper method of philosophy mathematical and emulate the way sciences represented their truths also in philosophy. But the problem of rationalism was even more unbearable than the Humean scepticism: a stagnation in problematical conceptions such as God, soul and freedom of the will, which were all facing proliferating critique throughout the 17th and 18th centuries. The result of this development was a multitude of attempts to formulate new philosophical systems which might overcome or erase the new wave of scepticism and secure direct human access to the surrounding reality. In this defensive battle the concept of intuition received a host of varying interpretations.

---

[11] LOCKE 1995, 433

[12] HUME 1999, IV:1,1.

[13] HUME 1999, IV:1,21.

### *3. Naturalistic intuition*

One of the most influential philosophical approaches of the 18th century was the Scottish Enlightenment and the common sense approach associated with it. Representatives of the Scottish Enlightenment, such as Adam Ferguson, James Beattie and Dugald Steward, valued immediate intuition-like sentiment as the proper human mode of knowledge. In contradistinction to Cartesianism, the Scottish understanding of the nature of intuition is clearly empirical. Intuition offers an unmediated sensate access to reality as it stands, and thus its tone is squarely opposite to that of idealism. The opposition to idealism is most apparent in the writings of Thomas Reid (1710-96), perhaps the greatest representative of Scottish philosophy.

According to Reid, the major philosophical trend of his time is something he calls "the ideal system". Reid traces the origin of the ideal system to Cartesian philosophy. The main principle of the ideal system is that whatever "is before the mind is either some impression or idea, or it is something that is represented by an impression or idea."[14] He considers the philosophies of Malebranche, Locke, Berkeley and Hume as representing refined modifications of Cartesian themes while being, nevertheless, versions of the ideal system. Because the ideal philosophies emphasise the certainty of inner convictions, they are easily exposed to doubt. According to Reid, scepticism is like an annoying parasite following the idealistic philosopher everywhere in his speculations. This parasite does not bother the naturalist, nor the common peasant. The ideal system can yield only counter-intuitive philosophy. Berkeley's idealism is directly opposite to the dictates of common sense in its absurd denial of an external material world. In the same manner, Hume's ideal philosophy is counter-intuitive in its scepticism.[15]

The philosophy Reid proposes instead of the ideal system consists of two major features. Firstly, Reid discards the assumption of the human mind as *tabula rasa*. On the contrary, he asserts, the human mind must be equipped with certain innate conceptions, such as the concept of extension, the concept of body, the concept of power, and the concept of self. Reid claims that all of these conceptions are irresistable and universal. The second important feature of Reid's thought is empiricism. Reid reverses the

---

[14] Lehrer 1989,12.

[15] Reid 2000, 23.

Cartesian approach, in which the assumption of mental states paved the way for proving the reality of the external world. From Reid's point of view there is no reason to rely more on the evidence of reason than on the evidence of perception.

> Sir, why should I believe the faculty of reason more than that of perception; they came both out from the same shop, and were made by the same artist...[16]

Although Reid does not expressly use the term "intuition", he certainly approximates intuition in several instances. This is especially clear when he discusses the way the external world *suggests* itself. According to his view, sensations immediately suggest external objects to the mind. These suggestions of the five senses "are judgements of nature" and are "immediately inspired by our constitution".[17] An essential feature of natural suggestion is that it is impossible to raise any reasonable doubt against its testimony. When, for example, we experience change in our surroundings, this change "suggests to us the notion of a cause, and compels our belief of its existence."[18] Those wanting to dispute such natural suggestions are, according to Reid, nothing but sophists and madmen. They inflict havoc on the common use of language especially by inventing private connotations:

> But if he [the philosopher] puts a different meaning upon the word, without observing it himself, or giving warning to others; he abuses language, and disgraces philosophy, without doing any service to truth; as if a man should exchange the meaning of the words *daughter* and *cow*, and then endeavour to prove to his plain neighbour, that his cow is his daughter, and his daughter his cow.[19]

Reid saw clearly the connection between intuitive understanding and the everyday use of language. Quite independently of the Scot, German philosopher Johann Georg Hamann (1730-1788) arrived at a judgement of the same kind. Hamann's thought emphasises the nature of reality as Word spoken by God. The language spoken by God is not, however, abstract and philosophically refined gibberish, but the language of a poet and "[p]oetry is the *mother-tongue* of the human race."[20] For Hamann the intu-

---

[16] REID 2000, 169.
[17] REID 2000, 37.
[18] REID 2000, 38.
[19] REID 2000, 39.
[20] HAMANN 1950, 197.

itive power of "natural language" yields for us reality with a divine mandate. This mandate is far removed from the conceptions of rationalistic theologians who seek for logos fitting to parameters of "despotic" reason (single, eternal, immutable, universal). Quite the contrary, natural language is always a language of a certain people in a certain time and certain place.[21] As has been often noted, this assertion makes Hamann's philosophy somewhat paradoxical. Reality is at the same time absolute (divine) and relative (human). The only solution to this dilemma seems to be an approval of dualistic metaphysics.[22] All apprehension is seeing made possible by the divine light. The result is an intuitive access to reality, which in practise resembles the common sensical approach to the universe.

## *4. Intuition and Critical Philosophy*

The espousal of realism and the linguistic turn of 20th century philosophy are among the reasons that we perhaps appreciate the philosophies of Reid and Hamann more than many of their contemporaries did. In retrospect, it was peculiarly fatal that Kant adopted a critical stance towards naturalism.[23] From Kant's perspective, common sense philosophy is naively uncritical. Kant opposes Reid's approach by claiming the human cognition of reality to be much more complicated than the Scot believes. Kant calls the naturalistic approach to the universe "mere misology brought to principles".[24]

Kant's critical philosophy probably occasioned the most substantial change in the understanding of the concept of intuition (*Anschauung*) by giving it a defined technical usage. In his own way Kant also relates the scope of intuition to something immediately available. In *Critique of Pure Reason* we are told that whatever intuition might capture, it must be spatial and temporal. In themselves, the original representations of space and time are not concepts but given intuitions *a priori* which make outer and inner appearances (*Erscheinungen*) possible.[25] Space and time are, thus, intui-

[21] O'Flaherty 1966, 16.

[22] O'Flaherty 1966, 56-57.

[23] See, however, Manfred Kuehn's important work, *Scottish Common Sense in Germany, 1768-1800* (1987). See also Wolterstorff's *Thomas Reid and the Story of Epistemology* (2001).

[24] Kant 1998, A855/B883.

[25] See Kant 1998, B40; A30-32/B46-48.

tions of a very special kind. In other respects, intuition and sensation are sides of the same cognitive issue. Intuition is immediately related to the object, or as Kant puts it, the "immediate relation of the power of representation to an individual object".[26] Sensation on the other hand is this same relation from the object's direction to the subject. For our purposes here it is sufficient to say that intuition is the "capacity for being confronted by individual items of reality."[27] Questioning of the reality of the external world was never Kant's intention.

We must ask, however, what the scope of intuition is? Kant reduces the realm of intuition to outer and inner sensibility.[28] His thought erases the once so apparent bond between the conceptual truths of reason and intuition. What human intuition captures is not something necessary, universal or eternal. The proper object of intuition is always a singular ontological object of experience (an entity). A remarkable feature of Kant's approach is that he considers intuition polar to concept. Not a single concept is intuitively available to the human mind. This means that intuition alone can never yield substantial knowledge. Intuition is noise in the sensory channels. Unlike what Reid assumed, according to Kant we simply have no immediate conceptual knowledge of causality, for example. Kant maintains that knowledge through concepts is always discursive and thus he explicitly denies the possibility of conceptual intuitive human knowledge.[29] Knowledge can take place only when the active powers of understanding cast their cognitive form on the raw material of sense data. Kant argues that the activity of our active understanding is something spontaneous and unforced.

It has been suggested that what Kant ultimately presents is sort of a blurred idea of direct reference.[30] This interpretation would propose that Kant truly wants to regard reality as something we are immediately in touch with. This, however, is a dubious conclusion. It is true that Kant tries to hold on to transcendental idealism and empirical realism simultaneously. But it is very difficult to maintain that in the end the Kantian ap-

[26] KANT 1928, 282 (KANT 1998, 719-720, editors note 2). See also KANT 1998, A19/B33.

[27] BENNETT 1966, 57.

[28] Occasionally KANT seems to equate intuition and sensibility, see KANT 1998, A43/B60, B75/A51.

[29] This is made clear when KANT states (1998, A51/B75): "The understanding is not capable of intuiting anything, and the senses are not capable of thinking anything."

[30] For this discussion see, for example, HOWELL 1973.

proach to the universe would be immediate and intuitive. It may be that through intuition I truly am in touch with the real world, but because of the transcendental conditions of experience I will always represent reality in a discursive and phenomenal manner.

> We have therefore wanted to say that all our intuition is nothing but the representation of appearance; that the things that we intuit are not in themselves what we intuit them to be... What may be the case with objects in themselves and abstracted from all... receptivity or our sensibility remains entirely unknown to us.[31]

Kant's ultimate argument is that the reality is truly intuitively given, but it can be *taken* only discursively, as a co-product of sensibility and understanding. Intuition seems to be a necessary but insufficient ingredient of knowledge. The interplay between passive receptivity and active understanding is probably one of the subtlest topics in Kant's philosophy, and my main task here is not to evaluate how viable Kant's approach is. I will, however, discuss some critical remarks made by one of Kant's fiercest opponents, Friedrich Heinrich Jacobi.

## *5. Intuition and Realism*

Jacobi makes expressly clear that his philosophy is a withdrawal "from a philosophy that makes perfect scepticism a necessity."[32] Although he speaks of "a philosophy", his philosophy actually opposes a vast variety of philosophies. Philosophies that lead to scepticism are, in Jacobi's opinion, hostile towards intuition.

> Ever since Aristotle the growing tendency among the schools has always been to subordinate immediate knowledge to mediated cognition. The *faculty of perception* that originally grounds everything has been subordinated to the *faculty of reflection*, which is conditioned by abstraction – in other words the prototype has been subordinated to the ectype; or the essence to the word, and reason to understanding.[33]

If we take a look at the circumstances in which Jacobi developed his ideas, two thinkers appear to be more relevant opponents than the others,

---

[31] KANT 1998, A42/B59.
[32] JACOBI 1994a, 193.
[33] JACOBI 1994c, 541.

Spinoza and Kant (Fichte). Jacobi discredits them both not only for degrading the true significance of intuition, but also for putting forward ultimately counter-intuitive philosophies.

We noted earlier that Spinoza was among the philosophers who embraced the idea that intuition is the highest form of knowledge. On this score Jacobi credits Spinoza. However, the case of Spinoza is more complicated. According to Jacobi, Spinoza, like all rationalists, tries to convert the original operations of intuition to reflectively apprehended concepts. Spinoza speaks highly of intuition, although in actuality he does not use intuition: he reflects. What Spinoza's intuition captures, then, is not the immediately available reality, but fictional and abstract conceptualisation. We might express this in other way by saying that the intuitive operation of "the mind's eye" has been systematised to its highest form in Spinoza's monism. However, all intuition of the mind's eye is mere speculation without a basis in the genuinely intuitive realm of nature. In the end, the result of Spinozism is a completely counter-intuitive theory of substance as one and infinite.

From this critique of Spinoza it is quite simple to induce Jacobi's critique towards Kant's use of intuition. We noted that Kant clearly relates intuition to the realm of real, ontological entities. Kant's realism is, however, mutilated realism. From Jacobi's point of view, Kant subordinates intuition to the dictates of understanding. Intuitions without concepts are blind. Intuitive cognition must always be qualified by the subject's discursive understanding. But why speak of intuition, if intuition doesn't actually yield immediate information about the universe? In fact Kant's philosophy raises new walls which complicate the intuitive approach to reality. Jacobi evaluates Kant's approach: "What would it be, at bottom, but the life of an oyster?"[34]

According to Jacobi, Kant fails to see how sensate intuition is always permeated (saturated) by conceptual content. Jacobi, quite to the contrary, insists that even the most primitive intuition is always conceptually laden. He thinks that

> the concepts of reality, substance or *individuality*, corporeal extension, succession, and cause and effect... must be common to all finite, self-revelatory, beings; and... that these

[34] JACOBI 1994b, 297.

> concepts have their concept-independent object in *the things in themselves* – consequently they have a true, objective meaning.[35]

According to Jacobi, we need no artificial additions to our natural capability of seeing things correctly. Jacobi compares the conceptual devices invented by rationalist philosophers to spectacles that allure us to believe they could enhance our healthy, natural capacity to see. However, people having excellent eyesight should not wear optics at all because their perceptions will become distorted. Instead, they should rely on their intuition. Intuition is a natural, unaided faculty of the human mind, which captures reality and, hence, is the only source of knowledge about objects. According to Jacobi there is, however, a mode of intuition that could be labelled the intuition of reason, "which gives us objects that transcend nature for our cognition, i.e. it makes us certain of their actuality and truth."[36] In rational intuition "something that the senses cannot reach is given as something truly objective, and not merely imaginary."[37]

Jacobi maintains that intuition is essentially passive. Intuition is always something occasioned, or as Jacobi puts it, revealed.[38] The revelation of reality also gives the mind an immediate awareness of most of the conceptual elements of the objective world.[39] Thus the active role of the human mind in a Kantian sense is diminished in Jacobi's thought. Human understanding can operate only with something already given. In the formation of original knowledge there is no need for the faculty of understanding, because the appearance of an object effects all our conceptual cognitions along with the merely sensible. Such knowledge is intuitive, for it is immediate and certain – at least as certain as human knowledge can ever be. Thus, we might say that Jacobi's theory of cognition aims at holism and simplicity. Jacobi argues that reality is something impossible to shut out. In order to underline the direct nature of its presence Jacobi sometimes designates this reality as "Thou". Reality interpreted as Thou and I emphasises intuitive and holistic experience, in which my existence and the existence

---

[35] JACOBI 1994b, 296.

[36] JACOBI 1994c, 563.

[37] JACOBI 1994c, 563.

[38] JACOBI 1994b, 276-277.

[39] Some conceptual features must be considered secondary or reflective and they result from rational comparison and evaluation. According to JACOBI this is the sole function of the discursive faculty of understanding.

of the external world is revealed at "the same indivisible moment".[40] Without "Thou" "I" would be dead; if "Thou" withdrew, nothingness would result. Here Jacobi's thought represents a surprising parallelism to Spinozistic metaphysics. Just like Spinoza, Jacobi also holds God to be the sole proper substance. Whatever the particular beings are, they are not substances – not at least in the rigorous sense of the conception according to which substances are those capable of existing independently of anything else. Human beings cannot sustain their existence without God. However, particular beings are neither mere modes of the one substance. Particular beings are persons. This means that they are individuals who can receive revelation or who have intuitive access to reality.

What Jacobi is tracing here seems to be some sort of combination of realism and fideism. When our intuition reveals something, we can only believe it. Nothing further can justify this belief. All Jacobi has to justify his realism by seems indeed to be something we might call blind faith, although Jacobi himself protested strongly against this line of interpretation. Jacobi frankly admits the impossibility of any other proof of an object's existence besides the very existence of the object itself. Circular as this reasoning seems, it is the only way of conceiving objects – unless we want to embrace idealistic metaphysics. Unwilling to take such a turn, Jacobi declares that the fact that we can become aware of any object in the first place must be considered inconceivable. Human intuition is miraculous.[41]

> The committed realist ... unquestionably accepts external things on the testimony of the senses. In keeping with this certainty he declines any other conviction; his only thought is that all concepts, even those that we call *a priori*, must have derived from this fundamental experience. What name, I ask, should such a committed realist give to the means through which he partakes of the certainty of external objects *qua* things existing independently of his representation of them? He has nothing to support his judgement except the fact itself – nothing but that the thing actually stand in front of him. Can he express himself in this regard with a more apt word than "revelation"?[42]

---

[40] JACOBI 1994b, 277, 288. Theological use of I and Thou is more subtle, yet crystal clear. Designating God as "Thou" is a practice undoubtedly adopted from pietism. God is essentially and primarily a person, not some sort of an abstract ideal of the rationalist philosophers. God is an ontologically indispensable being, without whose activity there would be mere nothingness.

[41] JACOBI 1994b, 272-273.

[42] JACOBI 1994b, 272.

## *6. Concluding Remarks*

What is the role of intuition in contemporary human approaches to the universe? At first glance it seems quite modest – at least in the sciences where intuition is probably best described as an auxiliary tool that might enable some important insights to pop up. Here the consensus ends. The problem of intuition is that it yields no systematic unity for building up a body of knowledge. As a methodological guiding thread in science intuition is even more dubious, because there is no one precise meaning or function designated for it and it seems to lead into some sort of privatistic notion of knowledge. For instance, we might say we are having an intuition of reality when we refer to the most intimate kind of awareness of our surroundings, as inexplicable and mysterious as this may be. Perhaps intuition is merely a hunch of some kind? Might the intuitive idea express merely undecided and uncertain cognition? Depending on the context, intuition could be characterised as occasioned, primitive, instinctive, innate, spontaneous or unconscious apprehension. The catchword of the scientific approach is critique – and intuition is never critical. It is, nevertheless, true that intuition may have some heuristic utility in the development of scientific theories. It might even be the case that the most significant discoveries and innovations of the future are basically intuitive. But it seems that since Copernicus the scientific approach to the universe has produced more and more results that seem to be counter-intuitive. In some cases it is of course true that the initially most counter-intuitive theories have turned out to be crystal clear, which this is often hailed as a true characteristic of the scientific approach.

Even today, however, when approaching the universe, we probably rely on intuitive sentiment more often than on any other cognitive power we possess. Intuition is genuinely a human approach to the universe. If this is a true supposition, it does not follow that such a human condition should be accepted at face value. Because intuition seems to be a trusting, even childish, approach towards reality, intuition can be considered applicable only to trivial everyday activities. When we must make mature judgements on serious life questions, or when the issue is more "academically tuned", we may underline methodological matters and critical investigation. If this is an apt account of the contemporary status of our use of intuition, then it is also true that we have given up the original prestige given to intuition.

The challenge for the sciences is to make its truths as plain and common as possible without straying into oversimplified populism. It is quite understandable that even during the most heated debates in science the hope of consensus and more unified opinion would not be forgotten. The straightforward naturalism of Reid and the naive realism of Jacobi could serve contemporary scientists as reminders to distrust too abstractive thought. If the truths offered in science are totally counter-intuitive and completely without relevance to human life, scientific knowledge might simply seem just alienating. In some respects, science must of course follow its exclusive ambitions, but this should not lead to despising the public discussion that craves for critical, weighted and sober-minded insights.

## *Bibliography*

ST. AUGUSTINE

1981 *The Trinity*. Transl. by McKenna, Stephen. The Fathers of the Church, Vol. 45. Washington D. C.: The Catholic University of America Press.

BENNET, JONATHAN

1966 *Kant's Analytic*. Cambridge: Cambridge University Press.

DESCARTES, RENÉ

1985 "Rules for the Direction of our Native Intelligence" in *The Philosophical Writings of Descartes*. Transl. by Cottingham, J. & Stoothof, R. & Murdoch D. Cambridge: Cambridge University Press.

DI GIOVANNI, GEORGE

1992 "The First Twenty Years of Critique: The Spinoza Connection" in Guyer, Paul (ed.): *The Cambridge Companion to Kant*. Cambridge: Cambridge University Press.

HACKING, IAN

1986 "Leibniz and Descartes: Proof and Eternal Truths" in Kenny, Anthony (ed.): *Rationalism, Empiricism and Idealism*. Oxford: Clarendon Press.

HAMANN, JOHANN GEORG

1950 "Aesthetica in Nuce" in *Sämtliche Werke*, Vol. 2. Wien: Verlag Herder.

HOWELL, ROBERT

1973 "Intuition, Synthesis and Individuation in the Critique of Pure Reason" in *Noûs* 7. 207-232.

HUME, DAVID

1999 *An Enquiry Concerning Human Understanding*. Oxford: Oxford University Press.

Jacobi, Friedrich Heinrich
1994a "Concerning the Doctrine of Spinoza in Letters to Herr Moses Mendelssohn" in *The Main Philosophical Writings and the Novell* Allwill. Transl. by di Giovanni, George. Montreal: McGill-Queen's University Press.
1994b "David Hume" [1787 edition] in *The Main Philosophical Writings and the Novell* Allwill. Transl. by di Giovanni, George. Montreal: McGill-Queen's University Press.
1994c "David Hume" [1815 edition; also "Preface and also Introduction to the Author's Collected Philosophical Works"] in *The Main Philosophical Writings and the Novell* Allwill. Transl. by di Giovanni, George. Montreal: McGill-Queen's University Press.

Kant, Immanuel
1928 *Handschriftlicher Nachlaß* (Gesammelte Schriften, Vol. 18). Berlin: de Gruyter.
1998 *Critique of Pure Reason*. Transl. by Guyer, P. & Wood, A. W. Cambridge: Cambridge University Press.

Kuehn, Manfred
1987 *Scottish Common Sense in Germany, 1768-1800. A Contribution to the History of Critical Philosophy*. Montreal: McGill-Queen's Press.

Lehrer, Keith
1989 *Thomas Reid.*

Locke, John
1995 *An Essay Concerning Human Understanding*. Amherst: Prometheus Books.

O'Flaherty James C.
1966 *Unity and Language: A Study in the Philosophy of Johann Georg Hamann*. New York: AMS Press.

Popkin, Richard H.
1992 *The Third Force in Seventeenth-Century Thought*. Leiden: E. J. Brill.

Reid, Thomas
2000 *An Inquiry into the Human Mind on the Principles of Common Sense*. Edinburgh: Edinburgh University Press.

Spinoza, Baruch
1982 *The Ethics and Selected Letters*. Transl. by Shirley, Samuel. Indianapolis: Hackett.
1985 "The Emendation of the Intellect" in *The Collected Works of Spinoza*, Vol. I. Transl. by Curley, Edwin. Princepton (NJ): Princeton University Press.

Wolterstorff, Nicholas
2001 *Thomas Reid and the Story of Epistemology*. Cambridge: Cambridge University Press.

# Expanding Ontology – A Locus for Autonomous, Subjective States in the Quantum Framework

TARJA KALLIO-TAMMINEN

## 1. Introduction

The classical paradigm of science was based on the idea of an objective material world which obeys strict deterministic laws. This clear-cut starting point was fruitful for physics which studied the macroscopic material world, but did not leave many options open to philosophers who tried to understand the relationship between human consciousness and the lifeless world of matter. In spite of many efforts to overcome the psycho-physical problem, a serious gap or divergence between science and humanism has remained unbridged for centuries.

As science typically attributes all motion and outcomes to external causes and rejects any internal principle of change, it is now quite common for people to believe that their mental activities or inherent subjective states could in principle be included within a computer. Reductive materialism has become increasingly popular even if we do not really know how neural brain-states are connected to subjective mind-states.

The majority of modern philosophers have attempted to hold on to the autonomy and responsibility of human beings even if within the world-view of classical physics, which did not provide an objective foundation for mental states, it was difficult if not impossible to understand how mind, or any kind of free choice, might interfere with the predetermined course of nature. Many philosophers resorted to Dualism or Idealism. The fundamental question of how it is possible for mind to have a genuine effect on the physical domain is still one of the basic problems in the philosophy of mind and in cognitive science.

The struggle between science and humanism has continued unsettled for several centuries, and in my opinion cannot be solved. By saying this, I mean that it cannot be solved within the classical paradigm of science. Even though this paradigm, which is based on the ideas of Descartes and Newton, has provided guidelines for several centuries of scientific explanation, this does not imply that this traditional conception of reality is actually true.

In recent decades, the classical paradigm has been challenged in many ways. In my opinion, the most serious criticism has emerged from physics. Quantum mechanics, the most important theory in modern science, cannot be understood or interpreted within the traditional context. The discussions concerning interpretation that have now lasted for a century have made it quite obvious that some of the fundamental presuppositions of the classical paradigm are inadequate, and that we are in need of ontological modifications.

In the quantum framework, the traditional space-time description of classical physics is replaced by a more-abstract description of state. The new approach offers philosophy of the mind better options, since the existence of causally-efficacious subjective states does not mean that they should be totally reducible to material particles or brain-states.

To clarify the issue, I will next discuss in more detail the formation of the classical paradigm of science and its basic ideas, as well as the new features revealed by quantum theory. I will then move on to see how these changes can illuminate the relationship between human mentality and the visible material world. To a certain extent, it appears to me that the formation of reality depends on the states we maintain in ourselves.

## *2. The Classical Paradigm of Science*

The modern conception of reality was born at the turn of the modern era. While greatly influenced by classical physics, it has, as does every wide-ranging theory or framework, certain ontological and epistemological presuppositions that of necessity remain metaphysical. In fact, the basic ideas on which modern science has been built come from antiquity. It was Galileo who combined the mathematical approach of Pythagoras with the atomistic and materialistic world-view originating from Democritus and Epicurus.

In antiquity, the early natural philosophers provided a variety of models concerning reality. Plato and Aristotle, whose extensive philosophies are often considered to be superior to pre-Socratic ideas, did not accept the particle-mechanistic conception of nature defended by the Atomists. They turned their attention from inanimate nature to living beings and saw that their growth and development could not be understood on the basis of the Atomist's idea of 'dead' matter. In addition to primal material, some kind of organising form was also required.

Plato believed that the visible sense-world reflected the world of ideas, while Aristotle argued that natural development followed hidden potentialities immanent in actualised objects. For both men, reality was an organic whole that contained both soul and spirit. This approach was later adopted by the Christian world-view. For example, Aristotle's ideas on physics and natural philosophy became an integral part of the mediaeval way of thinking. His way of handling causes and motion was considered superior to that of the Atomists, and his ideas were dominant for more than a thousand years.

It was only at the beginning of the modern era that discussion concerning natural philosophy regained the level it attained in antiquity. Classical physics began by abandoning the Aristotelian concept of matter and returned to Democritus' idea of dead matter. As matter was supposed to consist of indestructible tiny particles that obeyed strict deterministic laws, mind, reason and all subjective phenomena had to be stripped away from objective reality. Descartes divided reality into two strict parts: *res extensa* and *res cogitans*. He proclaimed that "neither God nor any rational soul present in the world will ever disturb the ordinary course of nature in any way". Physics began to study the measurable world of extended objects, subjective phenomena were excluded.

There is no need to discuss further the details of this mechanistic and deterministic way of thinking. The world-view that prevails in our society is still very much based on the 'facts' that are dictated by classical physics. The scientific approach and the world-view created by classical physics is often described as atomistic, mechanistic, reductionistic, or deterministic. It is founded on Dualism, with a detached observer being able to provide an objective and complete description of all natural phenomena.

Although this kind of simple model concerning the whole of reality is quite restrictive, I think it is still thought of as basically true. Conventional ontological and epistemological ideas concerning objects, their characteristics and their relationships, as well as our own relationship to them, are well suited to classical physics. Even though quantum physics has demonstrated that classical physics has severe limitations, classical metaphysics still very much guides our thinking and our imagination.

Classical physics did not say anything about the human mind but it has, nevertheless, had a considerable influence upon our approach to mental phenomena. The great philosophers of the modern era were forced to take the classical paradigm of science seriously. It was generally believed that physics had found the right idea in regard to matter. For centuries, it was impossible to question either the new conception of matter or the impressive results obtained by science, even if a philosopher wanted to focus attention on the mind. As the commonly-accepted world-view did not admit any place for non-material entities, few options were left open to the philosophy of mind. In addition to Dualism, the basic alternatives were Materialism and Idealism, both of which are, in a way, tied to the dualistic approach: everything must be reduced either to matter or mind, one or the other of Descartes' fundamental domains.

In scientific circles, Materialism has been popular since the Enlightenment, even though it was not accepted by many philosophers of the modern era. They also criticised Dualism. The great system builders such as Leibniz, Spinoza and Kant saw that in spite of the fact that he or she could not be taken outside it, a sensitive, willing and knowing human being could not be simply reduced to mere matter. Spinoza and Leibniz tried to create an overall picture of the world that contained a real spiritual or mental aspect in accordance with the material world, but encountered serious problems because of the commonly-accepted idea of matter (i.e. the classical paradigm of science). In his Panteism, Spinoza united mind and matter, but deterministic physics led him to deny free will. Leibniz proposed Parallelism, but was ultimately led to Idealism in order to guarantee subjective autonomy. Kant reached the conclusion that natural science was not able to deal with any speculative questions concerning, for example, ethics and metaphysics.

In philosophy, discussion concerning the possibility of free will and its effects has culminated in the confrontation between Incompatibilists and Compatibilists. These two groups take different positions as regards whether free will can be reconciled with deterministic nature. Compatibilists have traditionally assumed that practical freedom and responsibility are combatible with Determinism, whereas for scientifically-orientated thinkers it is usually quite evident that Determinism prohibits freedom.[1] The determinism of classical physics poses insurmountable problems for free will, because alternative possibilities of action require contingency. Since every action is determined by events that have already taken place, there can be no free agents in an overall deterministic framework.

Even if I sympathise with the Compatibilists concept of freedom, I do not agree with the Kantian idea of cutting back the domain of science. Freedom should be just the existence of genuine options and opportunities for action (and being able to choose between them), but it is not possible to totally disconnect mental or conscious phenomena from nature and natural laws. The clash between Compatibilism and Incompatibilism has remained unhealed within the classical paradigm of science. In my opinion, the great philosophers of the modern era became - in a sense - trapped, misled by the mechanistic conception of matter. Mental phenomena cannot be united to the predetermined course of nature in a natural manner, but this does not mean that the laws covering our inner activities should forever remain beyond the methods of science.

The scientific approach to mind and the psycho-physical problem received some elucidation when cognitive science provided the computer metaphor in the middle of the twentieth century. Mind was viewed as software and matter was seen as computer hardware. In Functionalism, mind does not have to be reduced to matter even if this approach is often connected with Materialism. Functionalism can also be seen as compatible with Dualism. Mind-states do not need to be identified with brain-states even if they are somehow realised in them.[2] For example, Patricia Churchland has argued that what counts for mentality is not the material of the entity but its internal structure.[3]

---

[1] As usually understood, Determinism implies that at any point in time the future is already fixed and unique, with no possibility of alternative development. In such a framework it is quite natural to think that the behaviour of humans is also determined by external events beyond their control. As physical indeterminism asserts that there are at least some exceptions to precise determination, many physicists, such as Compton, have enthusiastically welcomed the new theory. It is no longer justifiable to use physical law as evidence against human freedom (POPPER 1972, 217-220).

[2] REVONSUO 2001, 58.

[3] CHURCHLAND 1988.

In spite of the considerable progress made in brain research and neuroscience, our understanding of the nature of the interaction or connection between so-called material and mental aspects of reality has not evolved to any great extent. We do not really know how neural brain-states are connected to subjective mind-states. Cognitive science still faces the same basic problem the one faced by Spinoza and Leibniz. The computer metaphor is tied to the mechanistic and deterministic paradigm of science, and the customary algorithmic or computational approach does not allow the human mind any real creativity or free will, regardless of whether it is placed in a Materialist or Dualist framework. Within the classical paradigm of science, if we do not accept Idealism, we are led to believe that all our mental activities can, in principle, be carried out within a computer.

## *3. The Quantum Framework*

In the 1920s, quantum mechanics entailed a radical shift of paradigm in physics. As Newtonian mechanics had done in its time, quantum mechanics now offered the same kind of broad and universally-applicable theoretical framework that was capable of leading modern research. Even though the interpretation of this abstract theory is still a matter for discussion, it is evident that many typical quantum characteristics such as superposition and non-locality, entanglement, or statistical predictions cannot be understood within the currently generally-accepted particle-mechanistic and deterministic framework of classical physics.

The founding fathers of the Copenhagen interpretation, who were also largely responsible for developing quantum theory, understood that the illusion of a totally-predictable and autonomous external world broke down at the microscopic level. Quantum mechanics did not allow for the possibility of making observations without reference to the observer or to the means of observation.[4] The Copenhagen interpretation implied a radical reconsideration of classical metaphysics as well as a re-evaluation of the role played by human beings. Niels Bohr, the primary architect of the Copenhagen interpretation, said that human beings are both observers and actors deeply embedded in the same reality that their theories attempt to describe.

[4] For example WERNER HEISENBERG stated that natural science describes nature as exposed to our method of questioning, HEISENBERG 1962, 81.

Bohr realised that both classical mechanics and the world-view accompanying it are constructions of the human mind. He did not, however, mean that the forming of a model of reality was somehow arbitrary or subjective. At the same time as quantum mechanics surpassed classical mechanics, it also provided tools for the better understanding of reality. The classical paradigm turned out to be applicable only in the macroscopic world when we observe phenomena that are not directly related to us. Contrary to all expectations, intricate investigations in the microscopic world showed restrictions in the ideas of divisibility and space-time description that are typical of classical physics. This led Bohr to a more general framework, the framework of complementarity.

In describing reality, Bohr placed his trust in the extensive mathematical theories that physics can offer, but his principle of complementarity states that we are not able to take a God's eye-view . Neither in our everyday experience, nor in our natural language, do we have concepts, descriptions or models able to catch the truth as such. Every honest trial, however, may pick up a part of the truth, and wider models or theories provide us with perspectives that can unite circumstances previously seen as distinct.

Bohr's complementarity is sometimes seen as either vague or paradoxical since it does not appear to follow customary 'either-or' logic. He simply accepts the fact that, depending on the context, microscopic objects must sometimes be described as waves and sometimes as particles. Within a particle-mechanistic world-view this kind of strange behaviour is not understandable, but according to Bohr, quantum theory went far beyond the customary domain of classical physics. The theory, its state-description, was no longer based on the positions and velocities of particles in space-time. Because of the abstract wave-function, there was a new kind of wholeness in nature that remained outside the scope of traditional space-time description. The simple and visualisable particle-mechanistic model proved to be only suitable for macroscopic objects.

Many of the unexpected characteristics of quantum mechanics can be traced back to the new description of state based on the complex wave-function. It is the most important term in the theory and the biggest problem in quantum discussions has been interpretation of this complex entity, something which can never be directly observed. Some scientists think it is nothing more than a mathematical tool or instrument suitable for predicting the actual outcomes we observe. Others argue that the wave-function refers to some kind of transcendental quantum level. In any case, the wave-function is responsible for the non-local and statistical constitution of quantum physics. With the new description of state, some kind of indivisibility, internal spontaneity and change appear to be unavoidable components of reality.

Niels Bohr was the first physicist who truly immersed human beings in the same system as the rest of the universe. In his interpretation, the detached observer of classical physics can be seen as a constructive evolutionary agent. Bohr's complementarity means a denial of strict Dualism between subject and object. Reality appears as an evolving entity that can be divided and described in different ways. This means that our knowledge, our values and our goals may also play a notable part in shaping natural processes. Human actions and choices may direct development in nature, and what we call our 'subjective' aspects may have their own role in the dynamic web of reality.

Although Bohr replaced the passive observer of classical physics with an active participant who is deeply interconnected with reality, he did not go into ontological details concerning the relationship between human beings and nature. In contrast to Bohr, Wolfgang Pauli, also a member of the Copenhagen group, created an ontological model. Pauli spoke of a psycho-physical reality in which ordering structures, or cosmic archetypes of the Jungian type, work and manifest themselves in both the material world and in our minds. David Bohm, another famous interpretor of quantum mechanics, spoke of implicate order and active information as forming all reality. In my opinion, the basic ideas in the models proposed by Pauli and Bohm are quite similar. And, not surprisingly, both models have aroused fervent enthusiasm and severe criticism.

These kinds of ontological models of necessity remain speculative, and we do not need to go into any confusing detail. The most important conclusion is that quantum mechanics has led many researchers to reconsider both the classical paradigm of science and the role of human observers. In such models, mind is supposed to be profoundly intertwined with reality. Both strict ontological dualism and reductive materialism are avoided. The idea of a psycho-physical reality thus gives credibility to Pantheism or a certain kind of Parallelism (emergent levels). In contrast to the situation in the classical paradigm of science, Pantheism can be reconciled with free will in a quantum framework. Leibniz would also have been able to defend his monads without having had to fall back on Idealism.

## *4. A New Emphasis on Form: Beyond the Particle-Mechanistic Model*

Because of the complex state-description which is closely connected to the idea of abstract quantum fields, the conception of matter in modern physics has changed radically. No longer can the mathematical theories of physics be interpreted as describing material particles moving in space-time, which in turn means that the particle-mechanistic model proposed by the ancient Atomists has failed. Physics has proved a long-held ontological idea to be false, and has at the same time taught us something fundamental concerning the metaphysical basis of reality.

But what exactly is it that we have learnt? Mathematical symbols and abstract physical theories do not, unfortunately, carry with them a direct and clear interpretation concerning the character of the reality behind these theoretical representations. Proposing some kind of philosophical interpretation is a tricky business since there is not even a consensus concerning the nature of mathematics. Is mathematics a language with which '*The Book of Nature*' is written, or is it some kind of abstract tool we have invented? Whatever the answer, the mathematical tradition of physics is based on the belief that something in nature corresponds to abstract mathematical representation. This provides strong support for arguing that the state-function, the fundamental entity in quantum mechanics, also catches something essential to our understanding of reality. We are led to ask what kind of reality state-functions might describe.

The quantum state-function appears to refer to something which does not directly show up in the familiar material world we observe. Its form and solution depend on the characteristics of the studied object as well as on the context, i.e. on the boundary conditions that prevail in different circumstances, but in principle, the quantum state offers the probabilities for all possible occurrences in the three-dimensional material world. Many physicists provisionally divide reality into two levels: 1) the complex quantum level, in which superpositions etc, are at work; and 2) the classical level, in which only some of the possibilities allowed by the quantum level may become actual.

This kind of approach recalls the ideas of Plato and Aristotle who, in contrast to the Atomists, emphasised that all happenings and change in the material world cannot be ultimately reduced to local causes acting between separate particles which persist in empty space. Plato and Aristotle stressed form, as does quantum mechanics. In addition to the localised particle aspect, a holistic wave aspect also belongs to reality and profoundly affects the formation of physical events. Using Plato's and Aristotle's terminology, wave-configurations can easily be identified as certain kind of forms or internal patterns which control the formation of matter and are manifested in it. Their refutation of Atomism has become a subject of renewed interest: nothing in the visible sense-world seems to be without matter, but, at the same time, matter appears to be more than just matter. Matter and form are closely connected, integrated.

The great thinkers of antiquity would undoubtedly have asked whether state-functions belong to the eternal world of ideas or whether they are some kind of universals which guide the unfolding of different types of things, namely the actualisation of hidden potential structures. These complex entities of modern physics are in some way reminiscent of Plato's "perfect" ideas – as even the best measurements only project some part of the information contained in quantum states into our world, they can only be manifested imperfectly in the visible sense-world. It is also quite obvious that we are not able to truly enter the "quantum world" any more than Plato thought a human being could enter the eternal world of ideas.

Plato is often said to have totally separated the eternal world of ideas from the ever-changing sense-world. Even if some people like to separate the classical and quantum levels, quantum theory shows that the quantum world is not isolated. There is a clear connection between the macroscopic world and the quantum level since every occurrence at the visible (material) level changes the structure of future possibilities. Aristotle rejected Plato's strict boundary between the two worlds and attributed the "play of possibilities" to a non-being not directly detectable but which was, however, potentially something, a certain *kind* of non-being.[5] With his idea of potentiality, Aristotle stayed firmly in this world, as does quantum theory. We have a single reality which contains various kinds of tendencies or dispositions which may themselves become real and detectable.

Moreover, as the Copenhagen group clearly saw, the world also contains us, not just as external observers but as evolutive agents who, by manipulating the boundary conditions, are capable of shaping and modifying the

[5] According to scientific realism, the non-detectable can exist. As Parmenides concluded, non-being cannot exist.

obtainable outcomes.[6] Because of us, some of the tendencies (or propensities) in nature may proceed in certain directions. By knowing the relevant state-function, scientists are able to predict the distribution of the different types of results which may come out of similar-looking experimental setups. Using such a tool, they are able to approach context-dependent and individual phenomena that were outside the range of the classical paradigm of science. While this deeper knowledge has given us more power to deal with natural processes, it has at the same time shown us that we are responsible parts of the whole, and also that we are dependent on it.

## *5. Towards a Scientific Description of Subjective States*

If some kind of invisible states which can be approached mathematically also belong to reality, the primary obstacle to a physical understanding of mental states disappears. As the Incombatibilists clearly saw, there is no place for freedom or creativity within a particle-mechanistic framework since no outside factor, nothing 'mind-like' which is outside particle-mechanistic reality can ever have an effect on the closed deterministic processes of nature.

In a quantum framework, however, it is very much easier to locate both mental phenomena and human beings since the framework naturally includes a kind of "outside factor", i.e. the complex quantum-state which gives the probabilities of all possible occurrences but does not directly show up in the visible material world. If we suppose that abstract laws similar to those controlling the formation of matter are also in charge of forming mental structures, the quantum approach might offer an opportunity to bridge the gap between science and humanism.

The formation of our mental contents and structures is clearly connected with the plastic formation of neuron connections within the brain. I believe it is reasonable to think that the probability of a particular mental state becoming actual in a person's mind depends on their psycho-physical structure, their surroundings, and on their previous history. In any given context, quantum formalism is superior to classical physics when investi-

---

[6] For example, Heisenberg said that this was something that Descartes could not have attained, and it made the sharp separation between the world and the I impossible (HEISENBERG 1962, 81). Bohr continuously stressed that human beings are both orservers and actors.

gating complex holistic systems. If we had sufficient information concerning all the possible contents that might be actualised in any person's mind, we could perhaps, in principle, one day be able to write down a complicated function that would predict the probabilities for various possible outcomes in different kinds of circumstances.

Within a quantum framework, it is not necessary for mental states to be identified with brain-states on a one-to-one basis. The mental states may be real and causally-relevant, but this does not mean that they have to be reduced to material processes in the external world in order to be approached in a scientific manner. Even if the mental states only show up on a certain material basis, the non-local and statistical character of the quantum framework does not demand that they be completely identified with brain-states. The rich content of our inner selves can be reconciled with the natural order without there being any need to limit or deny its unique character.

## *6. Conclusion*

The quantum approach may offer us a better opportunity than any previously available of dealing with mental phenomena. The same kind of universal laws that rule development in material entities may also coordinate formation of our mental structures. In this kind of quantum framework connection between the mental and physical domains is obvious. Both the psycho-physical problem and the dispute between the Compatibilists and Incompatibilists disappear.

Solution of the psycho-physical problem does not demand that we choose between Materialism or Idealism. Actually, there is no requirement for the mental and material domains to be so very different. The boundary between those aspects of reality that we are accustomed to call 'mental' and 'material' need not be strict. Bohr maintained that the terms 'mind' and 'matter' are more or less just descriptive concepts that we use to understand the world we inhabit: they are not able to reach "transcendental" reality in as deep a manner as the best theories of physics. These mathematical theories do not tell us whether reality is ultimately made up of matter or spirit or something else. When trying to understand the nature of reality, we can only present hypotheses and gradually learn something from our mistakes.

Quantum theory has shown us that we do not live in a particle-mechanistic clockwork reality. The world includes many degrees of freedom and the past does not determine a unique future. Statistical laws as such do not offer humans any real freedom or subjective autonomity, but because of our genuine internal states, science is now also able to accommodate autonomous human beings. There is no need for us to be reduced to machines: the subject need not be an entirely-helpless victim of external causes.

To cope with the "inevitable course of natural events", whether they be external or internal, deterministic or statistical, the most important thing is to notice what is going on. Through knowledge, we can interfere with natural phenomena. In addition to purely material causes, both our purposes and our intentions can be seen as formative forces in nature. If we are able to affect our own mental states and act accordingly (i.e. if we can make evaluations and choices, widen our horizons, and relinquish states and habits that have proved to be harmful) something coming from within us, an inner conscious factor, does have a genuine opportunity to influence nature. I would say that anyone who takes the trouble to work in harmony with states which they consider to be more appropriate can be called an autonomous subject.

## *Bibliography*

BOHR, NIELS
1958 *Atomic Physics and the Human Knowledge.* New York: John Wiley & Sons.

CHURCHLAND, PATRICIA
1988 *Matter and Consciousness.* Cambridge (Mass.): The MIT Press.

HEISENBERG, WERNER
1962 "The Development of the Interpretation of the Quantum Theory" in W. Pauli (ed.) *Niels Bohr and the Development of Physics.* Oxford: Pergamon Press.

KALLIO-TAMMINEN, TARJA
2004 *Quantum Metaphysics. The Role of Human Beings within the Paradigms of Classical and Quantum Physics.* Doctoral Dissertation, University of Helsinki.

POPPER, KARL
1972 *Objective Knowledge. An Evolutionary Approach.* Oxford: Clarendon Press.

REVONSUO, ANTTI
2001 "Kognitiotieteen filosofiaa" in P. Saariluoma, M. Kamppinen & A. Hautamäki (eds.) *Moderni kognitiotiede.* Helsinki: Gaudeamus.

# Matter and Myths of the Soul: Perspectives on Plato, Carl Gustav Jung and Wolfgang Pauli

INGA R. GAMMEL

## *Introduction*

In 1954, the physicist and Nobel Prize Winner Wolfgang Pauli published an article concerning matter in which he addressed the psychophysical problem in contemporary Science as follows: "We have confidence that the human spirit will always be able to create ideas fitting somehow to the external objects that are conveyed to us by our sense data as witnesses of matter or energy in the sense of physics, because both the human spirit inside and the perceived object outside are subjugated to the same cosmic order."[1]

In these few lines Pauli touches upon a number of key issues essential to physics as well as to the age old conundrums in natural philosophy: What is matter, what is soul and spirit, and further, how are these phenomena related? And finally, how are we to understand the grand cosmic order? Since the day of the atomists we have progressed in the knowledge of matter, but according to present knowledge in the field of microphysics we actually still do not know what matter is. As to the concepts of spirit and soul, or psyche, we are on even shakier grounds. If we take our present understanding of, for example, soul it may be characterized as inferior to the conceptualizations offered by Antiquity. Modernity has substituted the concept of soul with the concept of mind.[2] Mind has come into the forefront, while soul is left behind as some sort of poetic *extravaganca.* Hence, soul has become something which presently is addressed only in pop songs and rock

---

[1] "Matter" in PAULI 1994, 34.

[2] As noticed by Paul Davies, even "the Catholic Encyclopedia settles for a definition of the soul as the 'source of thought activity'." See DAVIES 1990, 78.

lyrics. In brief, matter has lost its essence, and soul has become a non-question.

Until the seventeenth century matter was still thought to be imbued with soul. But with the emergence of modern science the physical-psychic problem disappeared from the Sciences, leaving the old idea of *Anima Mundi* as just a historical relic. How could it happen, asks Pauli, that such a great idea simply disappeared? And, how could it happen that the Sciences bowed down to such a narrow concept of reality?

With his bold claim that matter and psyche basically are manifestations of the same cosmic order, Pauli opposed the present concept of reality. After his sudden death in 1958, it came as a surprise to many of his colleagues and to the public in general that Pauli had indeed for many years been deeply engaged in developing a more adequate concept of reality that could also embrace the reality of the human soul. To illuminate what is at stake here, Pauli did speculate, from time to time, that the Cartesian *Cogito, ergo sum* should indeed be replaced with *Amo, ergo sum.*

Pauli did not arrive at such ideas out of the blue. In 1929 and 1930 he went through a deep life crisis, an event which took him into Jungian circles, and further, into a lifelong friendship and co-operation with the founding father of analytical therapy, Carl Gustav Jung. Here, Pauli was introduced not only to psychological issues such as feelings and emotions, but his intellectual horizon was also widened by philosophical, religious, and mythological issues. On the other hand, Jung, who, in his own words, had a poor knowledge of physics, benefited from Pauli's sharp criticism of matters in physics. Hence, the background of this interdisciplinary co-operation was mutual interest in how to develop a more adequate concept of reality. Over the years Pauli and Jung would meet for long hours of discussions, and they also exchanged their views in a comprehensive correspondence. In addition, Pauli also corresponded extensively with other colleagues and friends.[3]

It is indeed astonishing to learn that the efforts to revive the concepts of soul and of ensouled matter were launched, not from the humanities, nor

---

[3] In the years 1979-1999 Springer-Verlag published Pauli's scientific correspondence of 1919-58. References to Pauli's correspondence, with the exception of his correspondence with Carl Gustav Jung, refer to the numbers marked in brackets in WOLFGANG PAULI, Scientific Correspondence with Bohr, Einstein, Heisenberg a.o. Ed. by A. Hermann, K. v. Meyenn and V. F. Weisskopf. Berlin, Heidelberg, New York.

from theology, but from the sciences. All his life Jung had stressed that depth psychology is an empirical science, founded upon the experience of the psyche, of soul. Pauli, for his part, was a hard-core physicist who, because of a life crisis and existential sufferings, became engaged in issues outside his own narrow field of microphysics.

In this article I shall outline some perspectives on this mutual interest shared by Pauli and Jung. In different ways they tried to reshape our concept of soul and its link to matter. However, in a Western context any prominent theorizing on soul does, one way or the other, echo the master in this matter, namely, Plato. And the thoughts of Pauli and Jung are no exception to this rule. Therefore, I shall begin with some brief comments on Plato's cosmology, and his perception of matter and soul.

## *Plato*

First a general remark on Platonic scholarship. For centuries the reception of Plato has suffered from an unbalanced interpretation of the dualism at stake in his philosophy. A dualistic world view certainly is a characteristic of his early dialogues, but if we examine Plato's cosmology elaborated on in the dialogue *Timaeues,* a different picture emerges. Plato lived to become an old man, and though he, more or less, was engaged in the same questions for a lifetime, he did indeed, from time to time, change his opinions. As I shall argue, Plato's cosmology is a spectacular attempt to embrace matter and soul within a holistic world view.

In addition, and somewhat linked to the neglect of Plato's holistic attempt, we find another problem, namely, the modern devaluation of myth. Because of our insensibility or even contempt for mythology, some interesting information in Plato's cosmology, for example, of the nature of the soul, usually escapes attention. In the following, we shall take a closer look at these two problems.

### Plato's Cosmology

Plato's cosmology is the story of beginnings, but not of the Beginning. When the Demiurge, the Platonic creator daemon, shows up in the tale of creation, a number of axiomatic conditions are said to be already existing. Among those was pre-cosmic matter, but it was lingering in a state of disorientation and with no stable forms. Pre-cosmic matter reflected the eternal forms, but in a passive and incomplete way. Only when the Demiurge takes action, is this natural process guided in a specific direction. The main form, already being reflected by pre-cosmic matter, is believed to be the triangle. Thus Timaeus, the natural philosopher, after whom the dialogue has been titled, speculates that the Demiurge was just reinforcing an already initiated process by implementing more geometry. Besides triangles, geometric knowledge such as proportion is used to bind things together, to harmonize, and to create unions. In this fashion matter is stabilized in order to become building blocks in the creation of the visual universe, the body of the universe.

Thus the Demiurge is not engaged in the generation of the elements, but only in their perfection. Timaeus leaves his explanation open to the possibility of yet more profound principles known only to the Demiurge himself. But the fact that the Demiurge is not the maker of pre-cosmic matter may also indicate that he does not know the final principles. The Demiurge is not an omnipotent being like the Christian God. Having devoted himself to the realization of his vision, the Demiurge takes on the task of co-operation and in doing so he fortifies forces which have already revealed themselves as having a strong natural inborn tendency towards beautiful order. Generation in itself seems oriented towards order and beauty. By his operations along this line, imposing his own will and skills on the face of generation, the Demiurge does indeed introduce something new, that is to say accomplishment, perfection, permanence, and substantiality, which might otherwise not have emerged.

At this point the progression of the story is interrupted by the introduction of yet one more beginning, namely that of the World Soul. In comparison, matter was just brought into stable forms, but soul is a creation on a higher level. Soul is needed because order as such cannot stay ordered through its own quality. Hence, soul was in fact created before the creation of the body of the universe. The corporeal universe, it is said, was placed within the World Soul as she was the very first of created beings, and she was created in order to be "the mistress" of the body and to rule over it.

(Cf. *Tim.* 34 c)[4] Timaeus, the natural philosopher who unfolds the tale, cannot quite explain why he started with the description of the body of the universe and holds that it was caused by chance. However, this confusion in the progression of the tale echoes a specific feature of our physical world, namely its visibility. In contrast, soul is invisible, but her strong impressions can be experienced. Plato makes a great effort to elaborate on the imagery how the World Soul is interwoven into all substances of the universe, leaving nothing untouched by soul. In this way the universe was transformed from its previous chaotic state of being into a living animal. As is evident already from this brief sketch, matter and the World Soul are linked in a truly intimate relationship.

## Matter and the World Soul

Next, we shall focus on aspects of this delicate relationship. While pre-cosmic matter was simply brought into a more perfected state to become the elements to make up the body of the universe, the World Soul is created. According to this logic, the World Soul did not enjoy a pre-cosmic existence, but was created from some soulstuff, a mixture which is divided and bound together in due proportions. From Plato's text it is not clear what was actually put into the mixing bowl to make the World Soul. As it was left as a target for endless speculation on the origin and nature of this soulstuff, the creation of the World soul is shrouded in a real myth. How did she come into being? She obviously belongs to an order of her own. On the basis of Plato's text, vague hints can be established that some refined matter, in its most pure form, was used in creating a part of the World Soul.[5] Plato discusses different forms of fire, that is, light,[6] some types of which

[4] Quotations of Plato are from Plato, *Complete Works.* Ed. by John M. Cooper, Indianapolis 1997.

[5] When Plato describes the making of the human soul, the creation of the World Soul is indirectly illuminated. The Demiurge, it is said, "turned again to the mixing bowl he had used before, the one in which he had blended and mixed the soul of the universe. He began to pour into it what remained of the previous ingredients and to mix them in somewhat the same way, though these were no longer invariably and constantly pure, but of a second and third grade of purity." (*Tim.* 41 d).

[6] For example, it is said that the visible gods, i.e., the nature gods, are mostly made of fire "to be the brightest and fairest to the eye." (*Tim.* 40).

are so refined that they are invisible to the human eye.[7] Therefore, the assumption that soul may contain some refined matter and still be invisible does not embody a real problem. But I will have to leave this intriguing question here.

However, it is clearly stated by Plato that the World Soul was created in the shape of "an intermediate form of being". This means that the World Soul is a composite entity, participating in Being itself and at the same time familiar with the corporeal realm. Soul, then, has access to two worlds. (Cf. *Tim.* 35) She may, or may not, herself actually contain matter, but it is safe to claim that the text does emphasize that soul is always *in* matter. According to Plato's cosmology there is no such thing as dead matter as all matter is ensouled matter.

Hence, by her very nature the World Soul is destined to be linked with the universe, to be stirred, and to be moved. The same geometric principles by which pre-cosmic matter was ordered, are, more or less, also used to create the World Soul. Again, the link, or familiarity between Soul and matter is stressed. Thus Soul loves geometrical forms and harmonious order, both of which were used as the principles upon which she herself was generated. We may say that soul feeds on beauty, and beauty indeed is to be found in both worlds to which she has access, namely the world of reason and the world of visible beauty in the physical world.

## The Touch of Eschatology

Traditionally, soul is claimed to be immortal. Many of Plato's dialogues are in concord with this outlook. But in his cosmology another view is argued. According to the logic in this dialogue, everything created, or bound, can be decomposed, and this view does in fact touch upon some eschatological aspects. It is said that the creator daemon makes a Promise to his creatures that he will not draw back his creations. Hence, the Destiny of the universe dwells in the will of the Demiurge. If his will, for some unknown reason, disintegrates, or becomes subject to weakness or more powerful agencies, the body of the universe and the World Soul will decay together with its originator. This means that not only the existence of the visible universe but also the immortality of the soul is conditional. The eschatological

[7] Cf. *Tim.* 78.

stance in this logic further underwrites the claim that matter and soul are linked in unity.

However, Plato was adamant in his confidence that order and beauty would prevail. The threat of eschatology is kept at a distance by the Promise given by the Demiurge to his creatures: "Now while it is true that anything that is bound is liable to being undone, still, only one who is evil would consent to the undoing of what has been well fitted together and is in fine condition." (*Tim.* 41 b) It is an evil act to decompose what is beautifully made, and, therefore, argues Plato, the established order will remain. This argument, we may say, is basically founded only on aesthetic inclination. Thus Plato's cosmology is based on a genuine holistic view, embracing not only the world of eternal ideas, but also our world, the visible universe. It is stated several times that our kind of world *had* to come into being. Otherwise the universe would not have been complete. And such a viewpoint indeed originates in a deep experience of beauty.

## The Mythological Approach

Plato's cosmology is a unique blend of philosophy, scientific ambition, and mythic tales of how soul was created among other things. But to assume that the employment of mythic tales is a sign of inferiority certainly is an arrogance on the part of Modernity. Plato fully realized that human nature is constrained in terms of a complete understanding of the universe. Hence, cosmology can only be a "a likely account" (*Tim.* 53 d). Therefore, the use of myth provides a platform which allows the discussion to advance instead of running out of steam. The Greek approach to myth was not a simple one, but was rather sophisticated. As Socrates says: "No sensible man would insist that these things are as I have described them, but I think it is fitting for a man to risk the belief - " (*Phaedo* 114 d). Myth is no proper description and does not claim to be one.

Today, we don't really dare to be outspoken about the fact that we still must advance our knowledge by means of imagination. This was the problem to which Jung and Pauli devoted much energy and which I am going to discuss next.

## *Carl Gustav Jung: Psyche*

Whatever the opinions and critiques of Jung's theory of the archetypes, the collective unconscious, and his concept of psyche may be, his profound insights on the human soul are beyond dispute. Jung's theoretical work took its departure from practical experience. In his encounters with mentally ill and disturbed clients, Jung learned about the mysteries, the depths, the needs, and the sufferings of the soul. In addition, Jung himself went through a life crisis in which he experienced "the dark night of the soul", to use a phrase coined by the Spanish poet and mystic, Juan de la Cruz. Hence, to understand the nature of the soul became a life-and-death issue for Jung. Besides his existential and professional needs to explore the reality of soul Jung also found that the understanding of soul is desperately needed on a broader level. In general, Western culture suffers from a worldview in which soul is not accounted for.

### Soul and Myth

Before I sketch Jung's conceptualization of soul, it should be taken into account that a concept of soul, like the concept of any other basic entity, is provisional. In Plato's dialogue *Phaedrus,* Socrates wittily acknowledges this problem: "To describe what the soul actually is would require a very long account, altogether a task for a god in every way; but to say what it is like is humanly possible, and takes less time." (*Phdr.* 246) This was the wisdom of Antiquity, and it still prevails. In the same manner Jung emphasizes that we "cannot define 'nature' or 'psyche', but can only state what at present, we understand them to be." (*CW* 18, Par. 439)[8]

Therefore, in dealing with soul, even science must accept that the ultimate description of soul is only within the reach by employing myth. I want to underline this problem, because exactly in this matter Plato, Jung, and Pauli, as we shall see in a moment, all agree. Hence, across the centuries, and in spite of personal taste and professional interests the view still prevails that mythology and legend provide us with the most comprehen-

[8] The quotations of C. G. Jung are, if nothing else is stated, from the Bollingen Series. *The Collected Works of C. G. Jung.* Editors HERBERT READ ET AL. London and Henley (1957). Second Edition 1970.

sive and inclusive understanding of the nature of soul. No science, says Jung, will ever be able to substitute for myth, and a myth cannot be produced by science. Rather, that soul is shrouded in myth is a condition which should be taken into account. However, this fact should not interfere with the ambition to undertake a scientific research on the phenomenon of soul and her reality.

## Psyche as Nature

The fascinating and challenging aspect of Jung's view of soul is that soul is assumed to be natural. As such, soul is simply part of Nature. Like our bodies, the human psyche is also a product of evolution. And, accordingly, it shows many archaic traits as do our bodies. At a deeper level, man is still an archaic man. Many of our reactions testify to this familiarity with our predecessors. We are offended when people do not share our convictions, and we are insulted when somebody close to us finds our idea of beauty detestable. (Cf. CW 10, Par. 282) According to Jung we are animals, which we tend to forget, and therefore we neglect the primitive layers in our psyche. Soul does not originate in the present; her age is many millions of years old. Jung further assumes that our psyche, as part of Nature, is founded in accordance with the structure of the universe. Again, we hear some echoes from Plato, the repercussions of which are also present in Pauli's statement on matter and psyche, quoted in my introduction. Hence, what happens on a grand scale is being reflected in the life of the subjective soul who, so to speak, is adapted to our kind of universe. This Platonic viewpoint also provides the reason why we experience, for example, the circle, the square, and the four corners of the world with the cross as its axis as being of a numinous significance.

On the other hand, the idea that soul is adapted to the universe has implications as well. In a sense we are encapsulated in a certain psychic dimension. Our mode of observation does not necessarily grasp other modes of being, for instance, modes of being which are non-psychic. Here Jung points to a limitation in the condition of being human. But at the same time the psychic dimension is granted an optimal value. As Jung puts it, man's greatest instrument is his psyche. It is also "the greatest of all cosmic wonders and the sine qua non of the world as an object." (*CW* 8, Par. 357)

## Psyche between Nature and Spirit

Jung was convinced "that it (the psyche) is partly of a material nature."[9] Though the human psyche does show a historical development, Jung, nevertheless, found that she also embraces a dimension which transcends space and time. Thus the existence of the psyche is not entirely dependent on the corporal realm. In this sense the psyche is analogous to, or relates to, the unconscious and the land of the dead, which Jung speculated might be a condition, or level, in which "life" unfolds in a universe of images.

Thus Jung perceives soul almost in the fashion of Plato, specifically, as an intermediate being between matter and spirit. Jung says "The psyche as *triton eidos* and as a medium participates in both Spirit and Matter."[10] Hence, according to Jung the psychic is embedded or springs from a dimension which is non-psychic, that is, spirit. In his efforts to research the dimension of soul Jung came across the heights and the darkest depths of the human soul. He compared this overwhelming experience with the view of the night sky adorned with stars.[11]

While archaic man knew how to converse with his soul, we have become reluctant to allow ourselves to pay attention to the reality of soul. But even suppressed, the reality of soul carries through in our moods, in our choices, in our likes and dislikes, in the way we face our sorrows, in the way we enjoy and delight in things. And, finally the reality of soul also pushes through in our sense of the beautiful. As we shall deal with next, Pauli became engaged in showing how soul is also behind the choices which are being made by the scientist. The language, most akin to soul, is that of images, symbols, visions, dreams, imaginations, and synchronistic events. This part of Jungian theory fascinated Pauli greatly, and he made it *the touchstone* of his own ontology.

---

[9] THE PAULI / JUNG LETTERS 2001, 100

[10] THE PAULI / JUNG LETTERS 2001, 100.

[11] See "Introduction to W. Kranefeldt's 'Seccret Ways of the Mind'" in JUNG 1970, Vol. 4, par. 674.

## *Wolfgang Pauli: Symbolic Reality*

Pauli was greatly inspired by Jung's efforts to re-establish a unified world view, reviving the union between matter and soul. He truly shared Jung's ambition, but the outcome, the world view which Pauli elaborated on in his many letters, somehow even surpassed the ambition of Jungian theory.

From Jung, Pauli had learned how to read the symbolic language in his own dreams. Now, he projected his insight on the nature of dreams and their use of a symbolic language onto an ontological claim that reality as such does have a symbolic nature. To Pauli, reality at its deeper levels cannot be encountered. However, that said, we do indeed interact with an unseen reality through symbols. Symbols are our tools with which we engage ourselves in a dialogue with these unseen, deeper layers of reality. This view is already implicit in the Jungian theory which argues that all dealings with reality have a symbolic nature because they are psychic. But Pauli wanted to challenge this view, that is, to transgress the psychological view. I cannot in this context discuss the intriguing divergence between Jung and Pauli. However, it should be emphasized that this ambition drove Pauli to produce some powerful contributions to Jung's theory of synchronicity to which we shall return later. Next we shall take a look at some of the aspects of the Jungian theory which interested Pauli most and how he himself integrated this knowledge.

A theory claiming that we can only approach reality through symbols does indeed expose the problem of interpretation. How do we read the world of phenomena through symbols? The first question to be asked is, of course, what is a symbol? In an exchange of viewpoints with Aldous Huxley, who found the interest in symbols to be rather dubious,[12] Pauli states some important outlines. Symbols, symbolic images, and symbolic language are, so to speak, tools, by which we manage to receive new, or more adequate information about the obscure. Symbols display themselves on different levels of abstraction. The physicist and the mathematician are familiar with mathematical symbols, which are highly abstract models of reality. A more common level of symbolic language is encountered in art, but

---

[12] Huxley saw real mysticism, such as Zen Buddhism, as the most refined or sophisticated system because it has no use for symbols. For the discussion with Huxley, see Pauli to Huxley [2269] April 1956, Huxley to Pauli [2294] 10/6/1956, and, Pauli to Huxley [2322] 10/8/1956.

also in dreams, imaginations, visions, myths, and in synchronistic events.[13] We shall return to the latter in a moment.

Further, it is a characteristic of a true symbol that it cannot be invented, but that it is spontaneously revealed by the unconscious.[14] Pauli said that the symbol can be compared to a God who stirs the emotions of man but with the intention that man interacts or engages himself into a dialogue or an exchange with the symbol.[15] By doing so man gets into an interchange with an unseen reality. And he does so by means of interpreting the symbol and by materializing or transmuting its meaning to our level of reality.

Finally, it is crucial to emphasize that a symbol can never be fully exhausted in terms of interpretation. This problem has been addressed by Jungian scholar Marie-Louise von Frantz. According to her, symbols will "always retain an ineffable and mysterious quality that seems to reveal to us more than we can really know. If this were not the case, they would not be symbols but only signs or linguistic metaphors. The 'meaning' of a synchronistic incident only reveals itself on the condition that we understand the event symbolically and not just intellectually."[16] And she adds: "The realization of 'meaning' is therefore not a simple acquisition of information or of knowledge, but rather a living experience that touches the heart just as much as the mind."[17]

Jungian scholars have demonstrated that in dealing with the apparently irrational, there are rules to be followed. In consequence, von Frantz has examined the special laws for the connections between associations in myths and dreams. She says "the 'right' thought in the mythical context is the one that follows the 'necessary implications,' that is, the one that leads closest to the realization of the 'meaning'."[18] Hence, it is crucial to understand that an inherent logic is to be found in irrational events such as

[13] Pauli's claim that visions, imagination etc. are of a more primitive nature than mathematical symbols is, of course, correct in one sense. Yet in another sense a visual image may be all-embracing.

[14] See, for example, the discussion in von FRANZ 1992, 254 ff.

[15] For the full quotation, cf. Pauli to Fierz [971] 12/8/1948, in which Pauli writes as follows: "Das Symbol ist wie ein Gott, der auf den Menschen wirkt, der aber vom Menschen verlangt, dass er auf Ihn (das Gottessymbol) zurückwirken soll. Es scheint mir, dass nur so auch das Diesseits einen Sinn erhält und gerade das ist es, was mir bei Schopenhauer zu fehlen scheint."

[16] VON FRANTZ 1992, 254.

[17] VON FRANTZ 1992, 257.

[18] VON FRANTZ 1992, 260.

dreams and synchronistic events. Otherwise, we might slip into overinterpretation, or even back into magic.

All these aspects are integrated into Pauli's understanding of the symbolic nature of reality. Again, we become aware of the Platonic heritage which slips into Pauli's thinking through his adaptation to Jungian viewpoints. Plato advocated the legitimacy of myth. Now, Pauli, we may say, even radicalizes the poetic-aesthetic approach by claiming reality to be of a symbolic nature. In brief, Pauli framed his understanding around the belief that a symbol reflects an objective order in the cosmos. The phenomenon of synchronicity he saw as one of the most illuminating examples of how reality "works" in a symbolic manner.

## The Re-Establishment of a Union between Matter and Soul

To the surprise of everyone, the issue of soul became possible again, even in the scientific community, after quantum physics had stirred the materialistic concept of reality. Pauli was fascinated by the interface of matter and psyche, and when introduced to the theory of synchronicity by Jung,[19] he became enthusiastic about the prospects of developing a theory which might re-establish the broken union between matter and soul. He started working with Jung to develop the theory and contributed substantially to its final shaping.[20]

A synchronistic event is a falling together in time of an inner and outer situation but for which no causal cause can be established. For example, you dream that you are together with a person you have not seen in years. And the next day this person is actually standing at your door, or you receive a phone call or a letter. Or, you have for some time been thinking of changing your field of study into a certain direction. Then you walk into a bookstore to buy a newspaper, and the first thing to catch your eye is an important book to bolster your new interest. Such events have been richly recorded throughout history, and usually conceived of as signs of Divine

[19] The theory of synchronicity is closely linked with the theory of the archetypes. For discussions between Pauli and Jung on this topic, see Pauli / Jung letters 2001.

[20] Jung's article on synchronicity was published together with Pauli's essay on Johannes Kepler which dealt with the archetypal background of scientific thought and scientific concepts. See Jung & Pauli 1952.

Providence. This is the old idea of *kairos,* the right moment in time. But until Jung and Pauli took them up, these occurrences had not been subject to scientific investigation. Instead of using the word 'synchrone' Jung used the term 'synchronistic' in order to emphasize the special time aspect embedded in the phenomenon. It is not so that the two events must occur exactly at the same time, but rather that they may occur within a time span of hours, or days, but in such a fashion that the observer will remember his inner mood and therefore recognize the correspondence between his inner state and the outer event.

A further characteristic is that the phenomena of synchronicity are not of our choosing. We cannot make them occur. Instead they happen to us. Neither can they be reproduced, wherefore Pauli associates these phenomena as somewhat entangled with Destiny. The experience of a synchronistic event then is imbued with meaning and establishes a profound meaningfulness to the person who observes it.

A basic condition of becoming aware of such phenomena at all rests upon an excited state of feeling. This is the bottom line of the theory. If this psychic energy is missing, the chain of happenings may still occur, but the meaning, or the knowledge which was offered, is not grasped. Hence, it is crucial to understand that it is soul who experiences the phenomenon and makes it available to mind. While the emotionally charged atmosphere enables the synchronistic event to be recognized, mind is called upon to interpret the event. Only then does mind come into focus. Mind, or consciousness, is needed to interpret the occurrence that has been strongly experienced by soul.

Since the dimension of feeling is so crucial in order to understand symbolic meaning, it becomes obvious why Jung and Pauli adhered to traditional terms like soul and psyche.[21] Psyche is "the Greek term for soul, life and breath; so psyche *is* Nature itself."[22] While the concept of mind is a reductive concept, the traditional Platonic concept of soul refers to a comprehensive dimension that includes joy, buoyancy, humor, desire, love, passion, feeling, fascination, enthusiasm, pain, grief, sorrow, and rational thinking. Soul is all this and even more. Therefore a notion of soul be-

---

[21] In Plato's cosmology there is a hierarchial distinction between the World Soul as the more comprehensive in comparison to the human soul. Pauli wanted to discuss only the concept of the World Soul, which leaves some deficiencies in his theory.

[22] Cf. Jung 2002, 1.

comes a *conditio sine qua non* to a concept of symbolic reality. If soul is not acknowledged, then symbols, of which synchronicity is an example, cannot be detected and understood. With the synchronicity theory soul has regained her fundamental status in the context of philosophical, scientific, and cosmological speculation.

In his later years Pauli came to think, like Plato, that psyche is everywhere.[23] The phenomenon of synchronicity Pauli saw as a pointer to that direction. It is not possible, said Pauli, to determine the boundary where matter and psyche become separate. Pauli was inclined to see matter and soul as two different aspects of a higher principle, and since we cannot guarantee the non-existence of psyche in matter, we cannot uphold a waterproof gap between soul and matter. Therefore, according to Pauli, we must adhere to a unified theory. In short, we must adhere to a theory which allows the perspectives of the soul.

## *Conclusions*

In general we tend to think that demythification is an act of progression, guiding us towards a more sound perception of reality. And towards aesthetic inclination, which is favored by soul, we are rather suspicious. However, it must be taken into consideration that we can never fully escape the mythological nor our inborn propensity towards order, meaning, and beauty. In a bold manner, Jung, as well as Pauli, tried to reformulate and incorporate this knowledge in their theories. In my outline of Jungian theory I hope to have shown how Jung tried to make way for the reality of soul, an approach which was primarily characterized by a mythic-aesthetic vision of life. Pauli, for his part, aggrandized this aspect of Jungian theory and claimed that reality as such works in a symbolic manner. By these claims the idea of the World Soul, but also the idea of ensouled matter, once again became subjects of discussion. Cosmology today, be it philosophical or physical, still must pay attention to these ideas as the questions

[23] Cf. Pauli to Fierz [1507-Anlage] 19/1/1953: "Wo hört bei materiellen Phänomenen die Psyche wirklich ganz auf? Die Grenze wird sich niemals feststellen lassen. Und einmal hat auf dieser Erde ja auch jene Urzeugung stattgefunden, als zum ersten Mal ein grosses Molekül entstand mit einer 'hereditären' Reproduktionsfähigkeit. Wer kann die Abwesenheit von 'Psyche' dabei garantieren?"

and problems they embrace are still ours.[24]

Pauli was known as a sharp critic, frequently bantering with his colleagues that they put on their "thinking cap". Thus his sarcasms often had a witty edge. To illuminate the topics we have discussed in this context, the following anecdote may serve as an example: At one of the many Solvey congresses in Brussels, Pauli, Heisenberg, and Dirac were discussing religion in the hotel foyer. Heisenberg advocated that mankind has a religious need, while Dirac argued that religion is opium for the people. Pauli said nothing, but when Heisenberg asked for his opinion directly, Pauli said: "Now, I understand. There is no God, and Dirac is his prophet."[25]

## *Bibliography*

DAVIES, PAUL

1990 *God and the New Physics.* London: Penguin.

VON FRANZ, MARIE-LOUISE

1992 *Psyche & Matter.* Boston: Shambhala Publications.

JUNG, C. G.

1970 *The Collected Works of C. G. Jung.* Ed. by Read, Herbert & al. 2nd edition. London: Routledge & Kegan Paul.

2002 *The Earth has a Soul. The Nature Writings of C. G. Jung.* Ed. by Sabini, Meredith. Berkeley: North Atlantic Books.

JUNG, C. G. & PAULI, WOLFGANG

1952 *Naturerklärung und Psyche.* Zürich: Rascher.

PAULI, WOLFGANG

1979-99 *Scientific Correspondence with Bohr, Einstein, Heisenberg a. o.* Ed. by Hermann, A. & Meyenn, K. v. & Weisskopf, F. Berlin: Springer Verlag.

1994 *Writings on Physics and Philosophy.* Ed. by Enz, Charles P. & Meyenn, Karl von. Berlin: Springer Verlag.

2001 *Atom and Archetype. The Pauli/Jung Letters 1932-1958.* Ed. by. Meier, C. A. Princeton (NJ): Princeton University Press.

---

[24] A study of Pauli's conceptualization of symbolic reality will be forthcoming. In this I examine in more detail on the bonds between aesthetics, science, and mythology.

[25] TELEGDI 1988, 119.

Plato
1997 *Complete Works.* Ed. by. Cooper, John M. Indianapolis: Hackett Publishing.

Telegdi, Valentin L.
1988 "Pauli Anekdoten" in Pauli, Wolfgang: *Das Gewissen der Physik.* Ed. by Enz, Charles & Meyenn, Karl von. Braunschweig: Vieweg.

# Universal Order in Absolute Time

Tuomo Suntola

## *Introduction*

I like to start my introduction with a practical example of four clocks taken into different environment for one week.

We first have four identical clocks in a laboratory. They are observed to show same time at high accuracy for long periods of time. Then for one week we put one of the clocks to reduced temperature in a refrigerator, one clock into fast motion in a centrifuge, and one clock to an elevated altitude at the top of the laboratory building.

A layman states simply that the clocks in the refrigerator and in fast motion have run slower than the reference clock and the clock kept at an elevated altitude has run faster than the reference clock.

A physicist agrees with layman's conclusion regarding the clock in the refrigerator because he knows that temperature affects the ticking frequency of the clock...

...but there is no recognized physical mechanism for an atomic oscillator to change frequency due to motion – so he has concluded that time slows down for objects moving relative to the observer...

...there is no recognized mechanism for an oscillator to change frequency due to change in the gravitational potential either, accordingly, the reading of the clock taken to an elevated altitude can be explained if time is assumed to flow faster at a weaker gravitational interaction to a mass centre...

Regarding the clock in motion, the physicist's interpretation is based on the special theory of relativity originally created to explain the observed constancy of the phase velocity of light in moving frames. Regarding the clock at elevated altitude the physicist's interpretation relies on the general theory of relativity or equivalence principle which extends the conclusions of special relativity to gravitational potential through motion obtained in gravitational acceleration.

Are we unaware of possible physical mechanisms through which motion and gravitational potential affect the frequency of atomic oscillators or are we really obliged to blame time for the changed readings of clocks in motion or in different gravitational states? Does mass in whole space affect clocks in motion like it affects the acceleration of mass objects through inertia?

In his lectures on gravitation Richard Feynman stated:

> If now we compare this number [total gravitational energy $M_S^2G/R$] to the total rest energy of the universe, $M_S c^2$, lo and behold, we get the amazing result that $GM_S^2/R = M_S c^2$, so that the total energy of the universe is zero. — It is exciting to think that it costs nothing to create a new particle, since we can create it at the center of the universe where it will have a negative gravitational energy equal to $M_Sc^2$. — Why this should be so is one of the great mysteries—and therefore one of the important questions of physics. After all, what would be the use of studying physics if the mysteries were not the most important things to investigate.[1]

and further

> ...One intriguing suggestion is that the universe has a structure analogous to that of a spherical surface. If we move in any direction on such a surface, we never meet a boundary or end, yet the surface is bounded and finite. It might be that our three-dimensional space is such a thing, a tridimensional surface of a four sphere. The arrangement and distribution of galaxies in the world that we see would then be something analogous to a distribution of spots on a spherical ball.[2]

A closer study of Feynman's "intriguing suggestion of spherically closed space" leads to dynamic space described as a spherically closed structure expanding in the direction of the radius in the fourth dimension. Such solution shows the rest energy of matter as the energy of motion mass has due to the expansion of space in the fourth dimension. The dynamics of space is determined by the balance of motion and gravitation in the structure which explains the "the great mystery" of the zero-energy condition be-

---

[1] FEYNMAN, R., MORINGO, W. & WAGNER, W. 1995, 10.

[2] FEYNMAN, R., MORINGO, W. & WAGNER, W. 1995, 164.

tween gravitational energy and the rest energy of matter in space. Following the zero energy principle, any motion or gravitational state in space becomes related to the motion and gravitational state of whole space. Clocks in motion and clocks near mass centers run slower due to the linkage between the local energetic environment and the contribution of whole space. Contrary to the theory of relativity, in dynamic space time and distance can be handled as absolute coordinate quantities. We do not blame time for clocks running slow close to mass centers or moving at a high velocity.

Spherically closed dynamic space is studied in detail in the Dynamic Universe theory[3] ; predictions derived are supported by experiments equally or better than the corresponding predictions derived from the theory of relativity and standard cosmology model. The Dynamic Universe gives a holistic, highly ordered picture of space and universe. The multitude forms of local expressions of energy originate from and are related to the energy built up in the contraction - expansion process of spherically closed space. Mass appears as the substance for the expression of energy — mass as such is not observable, it becomes observable through momentum when in motion and through gravitation when at finite distance to other mass.

Dynamic space has generated the rest energy of matter against a release of gravitational energy in a contraction phase from infinity in the past to singularity turning the contraction to expansion. In the expansion phase the rest energy of matter is released back until zero at infinity. At infinity in the future, all motion gained from gravity in the contraction will have been returned back. Mass is conserved but it will no longer be observable because the rest energy of matter will have vanished along with the cessation of motion. The energy of gravitation will also become zero owing to the infinite distances. The cycle of observable physical existence begins at cessation in emptiness and ends at cessation in emptiness.

## *Gravitation and dynamics in spherically closed space*

The concept of space as a 4-sphere was first time suggested by Ludwig Schläfli and Georg Riemann in the mid-nineteenth century. In 1917 Al-

[3] SUNTOLA 2003.

bert Einstein proposed spherical space as a part of relativistic cosmology.[4] By that time, however, the general theory of relativity had just linked time to space as the fourth dimension, ruling out the possibility of the fourth dimension as the purely geometrical dimension required by an orthodox 4-sphere. Also, Einstein was looking for a static solution for space. Gravitation of mass within a 4-sphere would cause a shrinkage force, which would immediately throw the whole structure into a contracting motion along the radius in the fourth dimension. The reason for Einstein to suggest a cosmological constant was just to prevent such collapse. At that time understanding of distant space was vague, galactic structures were unknown; the recession of galaxies and expansion of space were discovered by Edvin Hubble about ten years after Einstein's concept of closed static space.

In spherically closed space a natural solution is not static space but space subject to contraction and expansion. Dynamics based on a zero-energy balance between the energies of motion and gravitation shows the rest energy of matter as the energy of motion mass has due to the contraction or expansion of space in the fourth dimension, in the direction of the 4-radius. This also means that the velocity of light is determined by and is equal to the velocity of space in the fourth dimension. In such a solution the velocity of light appears as the maximum velocity in space due to the energy balance in the system. In contraction started from the state of rest at infinity, in the past motion is gained against release of gravitational energy. In expansion motion works against gravitation resulting in gradual deceleration of expansion until rest at infinity, Figure 1.

[4] EINSTEIN 1917.

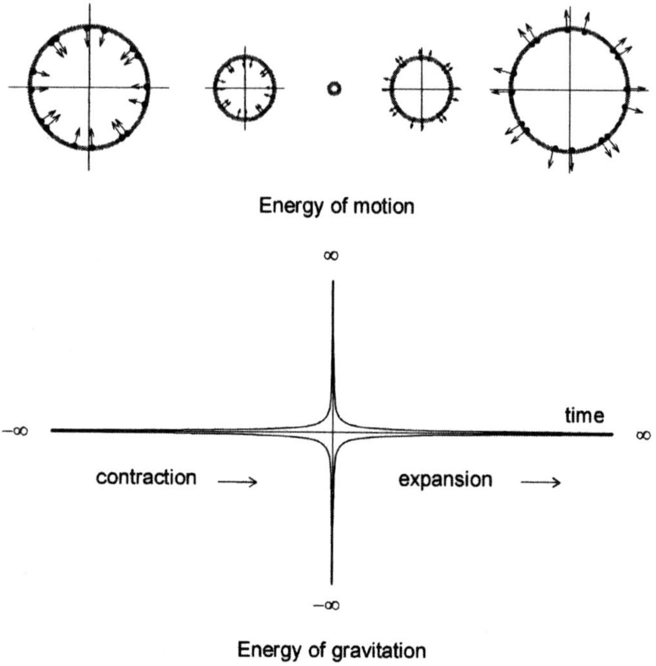


Figure 1. Contraction and expansion of space and the corresponding evolution of the energies of motion and gravitation. In the contraction phase, the 4-radius of space goes from infinity to zero. In the expansion phase, after singularity, the radius increases from zero back to infinity. Zero total energy is preserved through the entire process.

Motion of space in the direction of the 4-radius is undetectable for observers moving with the structure, indirectly it is observed as uniform expansion of space, the enlargening of all distances in space as the "three-dimensional surface" of the 4-sphere.

The kinetic energy related to the motion of space in the direction of the 4-radius appears as the rest energy of matter in space — being in balance with the gravitation of the structure, it discloses the mystery of the equality of the rest energy and the gravitational energy of all mass in space. The rest energy of any mass object is balanced by the gravitational energy of all the mass in space, i.e. the energetic state of a mass object is characterized as an excited state of the energies of motion and gravitation. It can be shown that the linkage of the rest energy to the gravitational energy of all mass in space also results in the inertial behavior of mass known as the Mach's principle.

Mass obtains the character of a substance for the expression of energy. Mass as such is not detectable, mass becomes detectable as matter with rest energy through motion and gravitational interaction with other mass in space.

## *Motion in four dimensions*

When the fourth dimension is interpreted as a geometrical dimension, as it should in a spherically closed dynamic space, a line element in the fourth dimension $|ds_4| = cdt$ means the distance traveled by space due to expansion in time interval $dt$. As a consequence, the momentum of mass $m$ due to expansion is $\mathbf{p}_4 = m\mathbf{c}_4$ and the energy of motion, when expressed like the energy of electromagnetic radiation propagating at velocity $c$, obtains the form $E_{m4} = c|\mathbf{p}_4| = c|m\mathbf{c}_4|$ or $E_{m4} = mc^2$, which shows the rest energy of matter as the "energy equivalence" of momentum in the fourth dimension. The total momentum of an object can now be expressed as the orthogonal sum of the momentum in space, in one of the three space directions, and momentum in the fourth dimension due to the motion of space as

$$\mathbf{p}_{tot} = \mathbf{p} + \mathbf{p}_4 \quad (1)$$

and the corresponding energy as

$$E_{tot} = c\,|\,\mathbf{p}_{tot}\,| = c\sqrt{|m\mathbf{c}_4|^2 + |\mathbf{p}|^2} \quad (2)$$

which is the well known expression of the total energy introduced by the theory of special relativity through a completely different reasoning.

The unavoidable momentum mass has in the fourth dimension due to the motion of space turns out to be the reason to the increase of the *effective mass* when mass objects are accelerated in space. As the counterpart of the increase of the effective mass of a mass object at motion in space, the *internal mass* working against the gravitation of all mass in space is reduced. Phenomenologically, this can be understood as an effect of the central acceleration created by motion in spherical space. For objects moving in space the energetic excitation (motion against gravitation) in the fourth dimension is reduced. It can be shown that the reduction of the internal excitation reduces internal processes in the moving objects. *Characteristic oscillation frequencies of atomic oscillators are directly proportional to the internal momentum of the oscillating object.* The internal momentum and the corresponding internal energy are

$$\left|\mathbf{p}_I\right| = m_I c = mc\sqrt{1-\beta^2} \qquad E_I = c\left|\mathbf{p}_I\right| = m_I c^2 = mc^2\sqrt{1-\beta^2} \tag{3}$$

where $m_I = m\sqrt{1-\beta^2}$ is referred to as the internal mass of the object. The concept of internal mass produces a reduced internal momentum and, accordingly, reduced characteristic frequencies for objects in motion. Clocks in motion run slower, time can be regarded as universally applicable coordinate quantity.

## *Local geometry of space*

The picture of regular spherical space is based on homogeneous distribution of mass as we can assume at cosmology level. Locally, mass has cumulated into mass centers. Following the zero-energy principle and the conservation of the total energy, it can be shown that cumulation of mass into mass centers in space result in bending of the 4-surface in the fourth dimension, Figure 2. In bended space the direction of local fourth dimension (local imaginary direction) deviates from the direction of the 4-radius. As a result the local imaginary velocity of space is reduced by factor cos$\phi$ from the velocity of the expansion in the direction of $R_4$, the direction of imaginary axis in homogeneous space, Figure 2. Reduction of the imaginary velocity of space means also reduction in the local velocity of light

$$c_\delta = c = c_{0\delta}\left(1-\delta\right) = c_{0\delta}\left(1-\frac{GM}{rc_0c}\right) \approx c_{0\delta}\left(1-\frac{GM}{rc^2}\right) \tag{4}$$

Bending of space together with a reduction in the velocity of light result in bending of light and delay of radio signals (the Shapiro effect) close to mass centers.

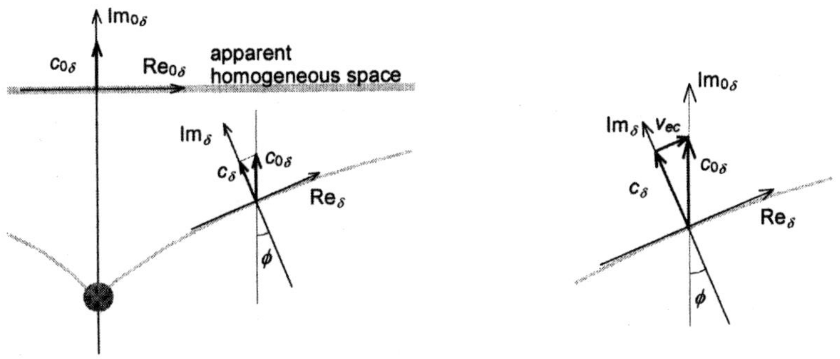


Figure 2. Space is tilted in the fourth dimension close to mass centers in space. Local complex coordinates follow the shape of space, causing the local imaginary axis, $\mathrm{Im}_\delta$, to deviate from the direction of the imaginary axis in apparent homogeneous space, $\mathrm{Im}_{0\delta}$. The imaginary velocity of space in the $\delta$-state, $c_\delta$, is reduced according to the formula $c_\delta = c_{0\delta}\cos\phi$, where $\phi$ is the tilting angle of space in the $\delta$-state. The orthogonal sum of the local imaginary velocity and the escape velocity, $v_{ec}$, to apparent homogeneous space is equal to the imaginary velocity of the apparent homogeneous space as illustrated in the picture on the right.

## *Characteristic oscillation frequencies*

We found that motion in space reduces the internal momentum of an object by reducing its internal mass. By reducing the imaginary velocity of space near mass centers, also a local mass center results in a reduction in the internal momentum of objects in the tilted space near the center.

It can be shown that the characteristic emission, absorption and oscillation frequencies of atomic objects are directly proportional to the internal momentum. As a result, the ticking frequencies of atomic clocks are reduced in direct proportion to the internal momentum. Taking into account the effect of each relevant energy frame of an oscillating object (like the gravitational potential and velocity in the Earth gravitational frame, the gravitational potential and velocity of the Earth in the Sun gravitational frame, gravitational potential and velocity of the solar system in the Milky Way, and of course the velocity of the clock in a local centrifuge frame as in our example of the four clocks we started with), the frequency determining the ticking frequency of an atomic clock is

$$f_{1,2} = f_{1,2(0)} \prod_{i=1\dots n} \left[ (1-\delta_i)\sqrt{1-\beta_i^2} \right] \tag{5}$$

where $f_{1,2(0)}$ is the frequency of the clock at rest in hypothetical homogeneous space. Equation (5) is in a complete agreement with all experiments made with atomic clocks on Earth, in satellites, in space crafts ect. For the derivation of equation (5) we have to observe that the Planck constant has the velocity of light as a "hidden" factor. In all circumstances the characteristic frequencies of atomic oscillators are in direct proportion to the velocity of light and the internal mass affected by motion as shown by the square root factor in equation (5).

The system of cascaded energy frames applied in equation (5) is a consequence of the zero energy principle: The energy of motion shall be paid back to the potential energy system it has been obtained from (or excited with); the kinetic energy of an object, the work done in achieving a velocity in space, is reversible, it is released when an object is returned to *the state of rest in the energy frame the motion has been obtained.* The rest energy of an object in hypothetical homogeneous space can be characterized as the intrinsic energy of the object, it is released only in billions and billions of years through the work the expansion of space does against gravitation in the direction of the 4-radius of space.

The linkage of kinetic and structural energies through zero energy principle gives an energy frame the character of an energy object with dualistic expression energy. Any local energy frame can be regarded as an energy object in its parent frame. Ultimate parent frame is hypothetical homogeneous space. The state of rest in hypothetical homogeneous space serves as a universal reference for any energy state in space.

## *A quantum of radiation*

We can find that the velocity of light is a hidden factor in Planck constant, accordingly, a proper way of writing the Planck constant is

$$h = h_0 c \tag{6}$$

where $h_0$ is referred to as the *intrinsic Planck constant* independent of the expansion of space. Applying the intrinsic Planck constant the energy of a quantum gets a unifying form

$$E_\lambda = \frac{h_0}{\lambda} c^2 = h_0 f \cdot c = p \cdot c = \frac{h_0}{\lambda} c \cdot c = m_\lambda c^2 \tag{7}$$

where $h_0/\lambda$ has the unit mass [kg] and can be characterized as the mass equivalence of an energy quantum (of radiation).

A point emitter like atom travels with space a distance of one wavelength in a cycle. When such emitter is described a dipole of one wavelength in the fourth dimension we find, just by applying Maxwell's equations, that the energy emitted in one cycle with only one unit charge oscillating in a dipole, is equal to

$$E_{\lambda 0} = 1.11 \cdot 2\pi^3 e^2 \mu_0 c \cdot f = h \cdot f = h_0 f \cdot c = \frac{h_0}{\lambda} c \cdot c = p \cdot c \tag{8}$$

Equation (8) give a physical meaning to energy quantum: *Quantum of electromagnetic radiation is the energy of one cycle of radiation emitted by one oscillation cycle of a unit charge in a quantum emitter.* The Heisenberg uncertainty principle turns into a clear message: *In order to obtain full information about the substance available in an energy quantum for the expression of momentum, we need to observe the full wavelength of the wave.*

## *Cosmological consequences*

Spherically closed space gives a more ordered picture of the universe than that suggested by the standard cosmology model based on the theory of rel-

---

[5] T. SUNTOLA, Appendix I published in www.sci.fi/~suntola/DU%20library/Universal%20Order%20in%20Absolute%20Time,%20summary.pdf.

ativity. The buildup and release of the rest energy of matter can be understood as a continuous process from infinity in the past through singularity to infinity in the future. The instantaneous singularity is not a "hot Big Bang" but an ultimate excitation of the rest energy of matter against the gravitation of the structure.

The contraction and expansion of space is considered as the primary energy buildup and release. As the initial condition the total mass of the universe is considered as being uniformly distributed throughout space. Condensation of matter into clusters in space is considered as a process of secondary energy buildup. The secondary energy buildup creates conditions, local singularities, "black holes", where the formless matter energized in the primary energy buildup is converted into electromagnetic radiation, elementary particles, and atomic structures.

In the preset state of expansion the 4-radius of the universe can be estimated to be about 14 billion light years (corresponding to Hubble constant $H_0$ = 70 (km/s)/Mpc). Since the expansion velocity has been faster in the past the corresponding time from singularity is 9.3 billion years. The expansion of space slows down gradually, present decrease in the expansion velocity (and the velocity of light) is $\Delta c/c \approx -3.6 \times 10^{-11}$ /year.

The expansion of space occurs uniformly everywhere in space, not only between galaxies or galaxy groups as taught by the standard model. The dimensions of galaxies and the radii of orbiting stellar systems are subject to expansion. Atomic radii, however, are conserved. The precise geometry, dynamics and energetic development of space allow the derivation of precise mathematical expressions for redshifts, optical distances, angular sizes, and apparent magnitudes of cosmological objects throughout the development of the Universe. The background radiation obtains a precise expression as the radiation propagated 360 degrees around expanding spherical space. What we see in the background radiation is the light emitted by the location of our own galaxy about 750 000 years after the singularity. The redshift of background radiation is $z = e^{2\pi} - 1 \approx 534.5$.

Predictions derived from the spherical, dynamic space in absolute time pass key cosmology tests equally or better that the corresponding predictions derived from the standard cosmology model (magnitude/redshift, Euclidean appearance of the angular size of distant objects, perihelion advance of elliptic orbits, bending of light and delay of radio signals near mass centers).

It is remarkable to notice that for a satisfactory fit with observations, the interpretation of recent supernova observations (magnitude/redshift) with standard model requires an assumption of "dark energy" with gravitational "push" at cosmological distances. Mathematically, dark energy is demonstrated by re-introducing a cosmology constant rejected by Einstein after Hubble's findings of expanding space in late 1920's. Such solution meant that the expansion of space were accelerating instead of decelerating as predicted by the "classical" Einstein - deSitter model.

The prediction derived from the Dynamic Universe (DU) theory gives an excellent fit with observations without any free parameter or new assumption.[5] It means that the expansion of space continues forever with a decreasing velocity until zero at infinity just as required by the zero energy condition between the energies of motion and gravitation in space.

## *Conclusions*

As a basic feature of scientific thinking, the reality behind natural phenomena should be understood unchangeable and independent of the models we use to describe it. The best a scientific model can give is a description which makes the reality understandable, rely on sound basic assumptions and internally coherent logic, and, specifically in physics and cosmology, gives precise predictions to phenomena observed and to be observed.

We can identify three kinds of qualities or factors a physical model is based on

1. *Basic laws of nature, fundamental quantities and natural constants*
   The identification of the laws of nature is based on experience and reasoning of the general "rules" nature is found to express itself.
2. *Phenomena to be described as consequences of the basic laws*
   A successful description of a phenomenon generates predictions for observations made or to be made.
3. *Coordinate quantities used as measures in describing phenomena*
   Coordinate quantities, the basic measures, allow quantitative expressions of physical phenomena in form consistent with human perception.

We are not free to choose the laws of nature but we have considerable freedom in choosing the coordinate quantities. Time and distance are the most fundamental coordinate quantities. For human perception and logic time and distance shall be universal for all physical phenomena described. It's a basic rule in all measurements not to change measures for a phenomenon in different environments or circumstances; adjusting the measures is a shortcut, a false way out of the problem.

There is no need to sacrifice the constancy and universality of the basic coordinate quantities, distance and time, for right predictions of the frequencies of atomic oscillators and the seeming constancy of the velocity of light.

*In space described as spherically closed entity with dynamics determined by zero-energy balance between motion and gravitation, absolute time and universal distance unit can be applied for all phenomena in space.*

The choice is this:

If we fix the velocity of light and make it the maximum speed in space by applying Lorentz transformation as a law of nature, we have to use time and distance as parameters in describing the behavior of atomic oscillators and measurements of the phase velocity of light in moving frames. We have to postulate the rest energy of matter as a property of mass and accept a relative state of rest as our reference for motion. We are not able to conclude either the overall geometry of space or the development of the geometry with the expansion of space.

If we describe space as a dynamic zero-energy surface of a 4-dimensional sphere, time and distance can be used as absolute coordinate quantities and we understand the dependence of the velocity of light on the development of the universe and on local gravitational state. We can relate the value of the velocity of light to the total mass and dimensions of space, we understand why and when the velocity of light is observed as being constant, and also why the velocity of light is the maximum speed achievable in space. Further, we can define the state of rest in any local energy frame and relate that to the absolute state of rest in space. We can identify a quantitative expression to Mach's principle, to the nature of mass, and an energy quantum of radiation.

Describing space as structured dynamic entity in absolute time gives simple and logical explanation to the mystery of the four clocks we started with; the clock in a centrifuge lost time because it was running slower due to its velocity in the centrifuge frame and the clock at elevated altitude gained time because it was running faster due to higher gravitational potential in the Earth gravitational frame.

The curious concept of time in theory of relativity was originally justified with and based on the relative velocity between the observer and the object. We know for sure, that the reading cumulated in the counter of a clock is unambiguously determined by the gravitational and motional state of the clock, it is not related to any particular observer. In fact, this was first time demonstrated in practice as late as in the 70's when atomic clocks were taken into airplanes and satellites. Until that time all experiments on the frequencies of atomic oscillators were based on observation of the frequency or wavelength of electromagnetic signal sent by the oscillators. Such experiments combine the inherent frequency of the transmitting oscillator, propagation of the electromagnetic radiation transferring the frequency, receiving of the radiation and the state of a reference oscillator at observer's location, which all can be handled in absolute time in dynamic space.

## *Bibliography*

FEYNMAN R., MORINIGO, W. & WAGNER, W.
1995 *Feynman Lectures on Gravitation* (during the academic year 1962-63). Reading: Addison-Wesley Publishing Company.

EINSTEIN, ALBERT
1917 *Kosmologische Betrachtungen zur allgemeinen Relativitätstheorie*. Sitzungsberichte der Preussischen Akad. d. Wissenschaften. Berlin: Verlag der Akademie die Wissenschaften.

SUNTOLA, TUOMO
2003 *The Dynamic Universe. A New Perspective on Space and Relativity*. Suntola Consulting Ltd.; www.sci.fi/~suntola.

# Einsteinian Time: The Microcosmic, Megacosmic and Mesocosmic

Raimo Lehti

We understand by *Microcosmos* the world of atomic and subatomic dimensions, by *Megacosmos* the world of galactic dimensions, and *Mesocosmos* as the world 'between'. Einstein developed concepts of time to explain phenomena in all three 'worlds'. Their distinctions may seem problematic as long as we take *time itself* as the fundamental concept. But Einstein took as his fundamental the concept of *clock*, thus time is defined as 'what the clock shows'. From that perspective, the distinct descriptions of microcosmic, mesocosmic and megacosmic *clocks* are quite plausible.

## *1. Classical concepts and its critics*

Before Einstein, *time* was regarded as the fundamental entity, and the *clock* was simply a device capable of measuring time. The principal text defining the classical concept of absolute time is found in Isaac Newton's *Philosophiæ Naturalis Principia Mathematica*:

> Absolute, true, and mathematical time, of itself, and from its own nature, flows equably without relation to anything external; … relative … time is some sensible and external ... measure of duration by the means of motion ….
> … The duration … of the existence of things remains the same, whether the motions are swift or slow, or none at all; and therefore this duration ought to be distinguished from what are only sensible measures thereof ….[1]

Newton's logic is, that the existence of absolute time is a prerequisite for

[1] Newton 1966, 6-8.

the success of temporal experiences such as those concerning the motions of celestial bodies. But those experiences do not provide the *meaning* of time; they provide methods for its approximate measurement. Clocks are more or less useful according to how correctly they measure the time. The *existence* of absolute time is guaranteed by the structure of mechanics which is built upon the use of the concept. The problem of time centered on the question of its *uniformity*; the *simultaneity* of distant events seems to have been taken as a non-problematic concept. Often one reads that when Newton accepted 'absolute' simultaneity, he assumed that signals with infinite velocity existed to communicate the time from one place to another. But no hint of such an idea is found in Newton's text, for he was quite conscious of the finite velocity of light. In his *Opticks* he writes:

> *Light is propagated from luminous Bodies in time, and spends about seven or eight Minutes of an Hour in passing from the Sun to the Earth.*
> This was observed first by *Roemer*, and then by others, by means of the Eclipses of the Satellites of *Jupiter.* ...[2]

The logic is clear. The moment of time for the *event* of the 'Eclipse of a Satellite of Jupiter' is something you learn *from the tables*, because they are calculated using a theory in which "absolute, true, and mathematical time" is the "equable" parameter of motions. The time for the *observation of the event* is not at all simultaneous with the event. The time of the event is not calculated from the time of the observation, though the time of the observation is *consistent* with the time of the event. After Newton, his definition of absolute time was reproduced by many of his followers. In 1884 James Thomson suggested that time does not of itself diffuse throughout all space but must be propagated from point to point by signals.[3] He wrote:

> Men have very good means of knowing in some cases, and of imagining in other cases, ... simultaneity at distant places. For this we commonly use signals by sound, by light, ... or by various other means.[4]

Thomson obviously thinks that the *existence* of simultaneity is not problematic, and the question concerns only human incapacity to ascertain it.

[2] NEWTON 1952, 277.
[3] TORRETTI 1996, 52.
[4] TORRETTI 1996, 294.

George Berkeley's *A Treatise Concerning the Principles of Human Knowledge* appeared in 1710. He takes 'time' to be something understood from the behaviour of human beings as 'clocks':

> Time, place, and motion, … every body knows; but having passed through the hands of a metaphysician, they become too abstract and fine, to be apprehended by men of ordinary sense. …
> Whenever I attempt to frame a simple idea of *time*, … participated by all beings, I am lost and embrangled in inextricable difficulties. I have no notion of it at all, ….[5]

Ernst Mach's criticism of the concepts of Newtonian mechanics was important for Einstein. In *Die Mechanik in Ihrer Entwicklung* Mach comments on the text of Newton:

> When we say a thing *A* changes with time, we mean simply that the conditions that determine a thing *A* depend on the conditions that determine another thing *B*.
>
> The vibrations of a pendulum take place *in time* when its excursion *depends* on the position of the earth.
>
> Since, however, in the observation of the pendulum, we are not under necessity of taking into account its dependence on the position of the earth, … the illusory notion easily arises that *all* the things with which we compare it are unessential. … Time, accordingly, appears to be some particular and independent thing, on the progress of which the position of the pendulum depends.... But … it is utterly beyond our power to *measure* the changes of things by *time*. Quite the contrary, time is an abstraction, at which we arrive by means of the changes of things …. [6]

Again 'absolute time' was taken to mean a 'uniformly flowing' time without reference to any motion of bodies. This time is not *relative* in the sense in which Einstein later defined the relativity of time. *Simultaneity* is not seen a problem; it was taken for granted. The 'Machian' criticism of the Newtonian absolute time was typical for many positivists. So Moritz Schlick writes:

> From the standpoint of the theory of knowledge, the objection was quite early raised against Newton, that there was no meaning in the terms Space and Time as used without 'reference to an object'; but, for the time being, physics had no cause to trouble about these questions ….[7]

---

[5] BERKELEY 1998, 138.

[6] MACH 1960, 272-273.

[7] SCHLICK 1963, 2.

## 2. *Albert Einstein: Mesocosmic clocks*

### 2.1. Time defined as the reading of a clock

Einstein presented the basic conceptions of the special theory of relativity in 1905 in the foundational article *Zur Elektrodynamik bewegter Körper*. The article begins with *Definition of Simultaneity*:

> It might appear possible to overcome all the difficulties attending the definition of "time" by substituting "the position of the small hand of my watch" for "time". And in fact such a definition is satisfactory when we are concerned with defining a time exclusively for the place where the watch is located; but it is no longer satisfactory when we have to … evaluate the times of events occurring at places remote from the watch.[8]

The insistence upon clocks as the primary concept instead of time "itself" raises a central question: what is a clock? When Einstein writes about the "small hand of my watch", he leads the reader to the idea of a *clock* as a rather commonplace mesocosmic device. Einstein then describes how watches located in different places can be synchronized using light signals, and continues:

> Thus with the help of certain imaginary physical experiments we have settled what is to be understood by synchronous clocks located at different places, and have evidently obtained a definition of "simultaneous", or "synchronous", and of "time". The "time" of an event is that which is given simultaneously with the event by a stationary clock located at the place of the event, this clock being synchronous, … with a specified stationary clock.

Einstein assumes the existence of a *species of clocks* having such a property that every one of them defines (in a stationary frame) the *same* time. He concentrates on the problem of *simultaneity* which for the 'Newtonians' was not problematic. Einstein takes for granted that 'time' needs a *definition* and not just a procedure to measure it. The positivists rejoiced: a fundamental concept was defined through an 'operative process' for its measurement. They often ignored Einstein's characterization of the experiments as *imaginary*; the experiments are often presented as real or realizable.

---

[8] EINSTEIN 1952a, 40.

The article *Über das Relativitätsprinzip und die aus demselben gezogenen Folgerungen* (On the Relativity Principle and the Conclusions drawn from it)[9] was published in 1908 in *Jahrbuch der Radioaktivität und Elektronik*. About clocks Einstein writes:

> For the temporal localization of a point-event we make use of a clock at rest relative to the coordinate system, adjacent to the point-event. The time of the point-event is defined as the simultaneous reading of the clock.
>
> We imagine an arrangement of clocks in several points at rest relative to the coordinate system. We suppose they all are similar, i.e. that the difference of the readings of two such clocks, when placed close to each other, shall remain constant. We also suppose the clocks to be set in some way; then the collection of clocks, when distributed densely enough, permits us the temporal judgment of an arbitrary point-event, for instance using the nearest clock.[10]

After the readings of the clocks at different points are compared using light signals, they can be used for a definition of time:

> The totality of the readings of clocks arranged as explained above, and understood to be at rest in the various points of the coordinate system, we call the time belonging to the coordinate system.[11]

For Einstein the fundamental concept is the *clock*, and *time* is defined as the common reading of a multitude of clocks arranged to be in concordance with each other. According to this choice of fundamental concept, the relativity of time and its dependence on the motion of the system looks like a quite plausible state of affairs. It asserts that the consistency of clock readings is defined in different ways in coordinate systems of differing motions. Hence philosophers cowed by the concept of the *relativity* of time did not direct their criticism at the moot point of Einstein's actual concept of time. They pictured a 'time' which could be bended, stressed and compressed, but such time did not exist at all in the special theory of relativity as understood by Einstein. Time as an independent ontological entity had disappeared; it was replaced by a computational scheme constructed using a certain kind of material object, the clock. According to this concept equations of motion were no longer equations expressing a relation between two 'existing' entities:

---

[9] EINSTEIN 1989, 433-488.

[10] EINSTEIN 1989, 437.

[11] EINSTEIN 1989, 438.

the position of some system and time. In principle the equations expressed a relation between two material objects: the moving system (for example a particle) and the hands of a clock. *This* was the point the critics of the Einsteinian concept of time should have concentrated upon. They did not; perhaps because the concept was so totally alien to people accustomed to the classical concept to time as something 'existing as such'. Einstein did not *change* the classical concept; in the special theory of relativity he *eliminated* it.

## 2.2. Attempts to define a clock

In his article *Le Principe de relativité et ses conséquences dans la physique moderne*[12], published in 1913, Einstein asserts that the hypotheses of special relativity are connected with the concept of time.[13] Everything is reduced to definitions using the readings of clocks. But what is a clock? We accept as a clock a material device characterized by phenomenon of returning periodically with similar phases; the 'principle of sufficient reason' compels us to assume that everything occurring in the device during one given period is identical with everything occurring during any period. If the clock is a mechanism having hands, the return of the hands to a given position allows us to count the number of periods that have passed. Measuring the time of the progress of past events means counting the number of periods of the clock between the beginning and the end of the progress. The definition must, however, be complemented when the clock is located far from the event.[14] To define the time for every space-point we must imagine the entire space filled with clocks of identical design. Their readings must be synchronized with the traditional method of light signals.[15] The totality of the readings of all clocks harmonized with each other we call *physical time*. In this way we define the time coordinates of elementary events as well as the simultaneity or not-simultaneity of two events.

Einstein wrote in 1913 about clocks:

> We are accustomed to measure time with clocks. Here we define a clock as a system automatically repeating exactly the same progress of events. The number of past events

[12] Einstein 1993, 131-176.
[13] Einstein 1993, 146.
[14] Einstein 1993, 147-148.
[15] Einstein 1993, 148-150.

> of this kind, counted beginning with an arbitrary one, we call the time reading of the clock, That time reading of the clock simultaneous with some event, we call the time of the event as measured by the clock.[16]

In a manuscript Einstein gives a similar definition of a clock:

> We consider a physical system, ... taking repeatedly some definite physical state *Z*. Then the state is always followed by states *Z'*, *Z"* etc., until the the system returns to state *Z*. The states of the system change periodically. Hence we may count how often the system which we will denote a 'clock', attains the state *Z*. This number we call the 'time mark' of the clock.[17]

Next we must be able to compare the time marks of two clocks remote from each other. This will be done in the customary way using light signals.[18] Again Einstein assumes the existence of a large family of similar clocks. The result of the procedure is:

> If after the completion of this procedure any two other clocks, for instance those located in points *A* and *B* are synchronized according to this definition, it finally does not make any difference that we have preferred the clock at *O* as that one used when all the others are synchronized. We can now characterize the result of the process as follows: every clock is synchronized with every other.[19]

This means that the synchronizing process creates an equivalence relation in the family of clocks, and the equivalence classes of this relation are 'times' of events (in the given coordinate system). This construction of time requires the existence of a vast family of everywhere evenly and densely distributed similar clocks. We might find this assumption almost as peculiar as the classical assumption of absolute time. The difference is only that the fundamental concept is a *clock* and not *time*. This change renders it almost a triviality that the 'time markings' are dependent on the space points where the clocks are located, and that automatically makes time dependent on a coordinate system:

> It is important that the imposing of a time coordinate and therefore also the simultaneity has a sense only when the coordinate system is given, with which the imposing is

[16] Einstein 1995, 541.
[17] Einstein 1995, 37-38.
[18] Einstein 1995, 38.
[19] Einstein 1995, 38.

> related. For measuring the time coordinate or for confirming the simultaneity we namely need a system of similar clocks arranged at rest relative to *K*.[20]

The assumption of the equivalence relation upon which the construction depends, is not at all a triviality. In his popular book of 1917 Einstein writes:

> We suppose further, that, when three events *A, B* and *C* occur in different places in such a manner that *A* is simultaneous with *B*, and *B* is simultaneous with *C* (simultaneous in the sense of the above definition), then the criterion for the simultaneity of the pair of events *A, C* is also satisfied. This assumption is a physical hypothesis about the law of propagation of light; ... [21]

## 2.3. The definition of simultaneity

A popular account *Über die spezielle und die allgemeine Relativitätstheorie* written by Einstein appeared in 1917. I use the English translation EINSTEIN 2001.[22] Einstein begins his analysis of time with the problem of simultaneity. Einstein gives his famous solution of this problem with an analysis of what meaning can be given to the statement that two lightning flashes occurring in different locations are simultaneous.

> After some time perhaps the following answer would occur to you: "The significance of the statement is clear in itself and needs no further explanation; of course it would require some consideration if I were to be commissioned to determine by observation whether in the actual case the two events took place simultaneously or not." I cannot be satisfied with this answer .... The concept does not exist for the physicist until he has the possibility of discovering whether or not it is fulfilled in an actual case. We thus require a definition of simultaneity such that this definition supplies us with the method by means of which, ... he can decide by experiment whether or not both the lightning strokes occurred simultaneously. As long as this requirement is not satisfied, I allow myself to be deceived as a physicist (and of course the same applies if I am not a physicist), when I imagine that I am able to attach a meaning to the statement of simultaneity. (I would ask the reader not to proceed farther until he is fully convinced on this point.)[23]

[20] EINSTEIN 1995, 38-39.

[21] EINSTEIN 2001, 25.

[22] EINSTEIN 2001.

[23] EINSTEIN 2001, 23-24.

This was an influential declaration. Einstein was not satisfied with the thesis upheld by Thomson that simultaneity *as such* is an unproblematic conception; the problem is only in its experimental verification in concrete cases. Does Einstein mean that it is a *general epistemological principle* that a "concept does not exist for the physicist until he has the possibility of discovering whether or not it is fulfilled in an actual case"? In this interpretation Einstein is presenting a very general positivistic or operationalistic philosophy, usable in all domains of knowledge. This was the interpretation many of Einstein's followers gave to his statement. Another interpretation is that Einstein did not have any idea of a general epistemological rule, but only a rule which should solve the *special problem of simultaneity* and nothing more. Einstein himself described later his own philosophy of science *opportunistic*, and that characterization would support the latter interpretation. Then epistemological considerations really did not play any role in Einstein's road to special relativity; he simply tried casuistically to solve the special problem of the concept of time haunting him. Einstein proceeds further:

> After thinking the matter over for some time you then offer the following suggestion with which to test simultaneity. By measuring along the rails, the connecting line *AB* should be measured up and an observer placed at the mid-point *N* of the distance *AB*. This observer should be supplied with the arrangement (...) which allows him visually to observe both places *A* and *B* at the same time. If the observer perceives the two flashes of lightning at the same time, they are simultaneous.
>
> ... I feel constrained to raise the following objection: "Your definition would certainly be right, if only I knew that the light by means of which the observer at *M* perceives the lightning flashes travels along the length $A \rightarrow M$ with the same velocity as along the length $B \rightarrow M$. But an examination of this supposition would only be possible if we already had at our disposal the means of measuring time. It would thus appear as though we were moving here in a logical circle."
>
> After further consideration ... you declare: "I maintain my previous definition nevertheless, because in reality it assumes absolutely nothing about light. There is only *one* demand to be made of the definition of simultaneity, namely that in every real case it must supply us with an empirical decision as to whether or not the conception that has to be defined is fulfilled.... That light requires the same time to traverse the path $A \rightarrow M$ as the path $B \rightarrow M$ is in reality neither a *supposition nor a hypothesis* about the physical nature of light, but a *stipulation* which I can make of my own freewill in order to arrive at a definition of simultaneity."[24]

[24] EINSTEIN 2001, 24-26.

In a footnote Einstein retracts this and makes a *supposition*: "when three events *A*, *B* and *C* occur in different places in such a manner that *A* is simultaneous with *B*, and *B* is simultaneous with *C* (simultaneous in the sense of the above definition), then the criterion for the simultaneity of the pair of events *A*, *C* is also satisfied." This he characterizes as "a physical hypothesis about the law of propagation of light". The text reproduced above concentrates on the question of simultaneity. Before Einstein the question of the 'absoluteness' of time concerned another problem, the 'uniform flow' of time. This question Einstein passes over somewhat nonchalantly supposing the existence of clocks which "go *at the same rate* if they are of identical construction".

## 2.4. The relativity of time

Einstein next shows that the definition given necessarily leads to a concept of the relativity of time, the most startling concept of his theory.[25] Einstein investigates what the definition implies for two observers, one on a stationary embankment and another on the moving train:

> Are two events (e.g. two strokes of lightning *A* and *B*) which are simultaneous *with reference to the railway embankment* also simultaneous *relatively to the train*? We shall show directly that the answer must be negative. …
>
> Events which are simultaneous with reference to the embankment are not simultaneous with reference to the train, and *vice versa* (relativity of simultaneity). Every reference-body (co-ordinate system) has its own particular time; unless we are told the reference-body to which the statement of time refers, there is no meaning in a statement of the time of an event.

The formulation: "Every reference-body (co-ordinate system) has its own particular time" is unfortunate, because it leads to the thought that time is something that *exists*, and hence there exists various times for different persons. One must go back to Einstein's definition of time to realize that according to his present view, the only *existing* things are clocks. When we use the formulation "Every reference-body has its own particular clock", the statement has no trace of a paradox (or even novelty). Another source of possible misunderstanding is the use of the word *observer*. The observer

[25] EINSTEIN 2001, 27-29.

is only a linguistic device. The use of this word has, however, given rise to the incorrect view that 'relativity' means some kind of subjectivity; a reference to the 'personal perspective' of the observing person. In the final formulation the 'observer' has been dropped out, and Einstein writes "Events which are simultaneous with reference to the embankment are not simultaneous with reference to the train ...". One observer is enough to register all that happens. Einstein's conclusion concerning the relativity of time was the most decisive thesis of relativity theory; at this point it deviated radically from the traditional physics. This was also the point on which the criticizers of relativity concentrated. When Einstein argues for the non-existence of absolute time using its incompatibility with his definition of simultaneity, this does not prevent the possibility that one of the time-parameters (i.e. a special clock) could for some reason be taken as a privileged one.

Later in 1936, in an article *Physik und Realität* Einstein writes:

> The illusion which prevailed prior to the enunciation of the theory of relativity – that, from the point of view of experience the meaning of simultaneity in relation to spatially distant events and, consequently, that the meaning of physical time is *a priori* clear – this illusion had its origin in the fact that in our everyday experience we can neglect the time of the propagation of light. We are accustomed on this account to fail to differentiate between "simultaneously seen" and "simultaneously happening"; and, as a result, the difference between time and local time is blurred.
>
> ... Such a use of notions – independent of the empirical basis to which they owe their existence – does not necessarily damage science. One may, however, easily be led into the error of believing that these notions, whose origin is forgotten, are logically necessary and therefore unalterable, ...[26]

Einstein somewhat nonchalantly accuses the prerelativistic physics of an illusion which is due to our failing "to differentiate between 'simultaneously seen' and 'simultaneously happening'". Certainly the astronomers at least since Newton (and Rømer) understood that we do not see events in the heavens at the moment they happen. Now Einstein backs away from the 'positivistic' accusations of the Newtonian time and admits that "Such a use of notions – independent of the empirical basis to which they owe their existence – does not necessarily damage science."

---

[26] EINSTEIN 1982, 299.

## *3. The relativity of Mesocosmic time before Einstein*

### 3.1. Lorentz and his transformation

In his investigations about different time-parameters to be used in physics Einstein had at least two precursors, Hendrik Antoon Lorentz and Jules Henri Poincaré. Lorentz gave another suggestion concerning what is a clock and what relation it has to time. His mathematical problem was to solve Maxwell's electromagnetic equations in a moving frame of reference. In 1895 Lorentz presented his results in a new way using a new and crucial equation concerning the *transformation of time*. It led Lorentz to introduce the terms *common time* and *local time*. In 1904 Lorentz published an article *Electromagnetic Phenomena in a System Moving with any Velocity less than that of Light,* which gives a good exposition of how to solve Maxwell's equations using the transformation later christened the *Lorentz-transformation*. Lorentz supposes that the system as a whole moves with a constant velocity *v* and writes the equations using the new coordinates including a new time coordinate *t'*.[27] Then he writes

> The variable *t'* may be called the "local time", indeed, ... it becomes identical with what I formerly denoted by this name.[28]

Lorentz, however, seems to have thought that there exists only one 'correct time'. In *The Theory of Electrons* Lorentz again introduces a new variable *t'*, which he calls "the *local* time, distinguishing it from the *universal* time *t*". He writes:

> It appears from what has been said that, by the introduction of the new variables, all the equations of the problem have again taken the form which they have when there is no translation. This at once leads to the following conclusion:
>
> If, in the system at rest there can exist a state of things in which [the field quantities] are certain functions of *x, y, z, t,* the moving system can be the seat of phenomena in which the [same] vectors are the same functions of the relative coordinates *x', y', z'* and the local time *t'*.[29]

[27] LORENTZ 1952b, 14; cf. LEHTI 2002, 146.

[28] LORENTZ 1952b, 15.

[29] LORENTZ 1952c, 189.

## 3.2. 'True' and 'effective' quantities

Lorentz introduces some concepts:

> ... Let us ... confine ourselves to the [moving] system S with which we are really concerned. We may introduce for this, as we have already done, the quantities $x'$, $y'$, $z'$, and we may even use them for the determination of the position of a point, .... Let them be called the *effective* coordinates, ...
>
> It is only for moving systems that we have had reason to distinguish between the effective coordinates and the "true" coordinates, the effective elements of volume and the "true" ones, etc.; as soon as $w = 0$, we shall have $x' = x$, ... etc. Yet, for the very reason of these equalities, we are free also to speak of the effective coordinates, the effective density, etc. in the case of a stationary system; only we must not forget that in this case these quantities are identical with the true coordinates, the true density etc. ...[30]

Lorentz begins to use the attribute 'effective' for quantities both in a stationary and a moving system and assures that according to this convention there is no difference in the electrostatics of the two systems (p. 201). Hence, the reader may arrive at the idea that there is little reason to assert some of these quantities as *true*. Lorentz, however, thinks that a difference between them exists and makes some remarks about the relation between his theory and that of Einstein:

> The denominations "effective coordinates", "effective time", etc., of which we have availed ourselves for the sake of facilitating our mode of expression, have prepared us for a very interesting interpretation of the above results, for which we are indebted to Einstein.[31]

Lorentz gives an exposition of how an observer in a stationary system and an observer moving uniformly compare their observations. He says concerning clocks:

> With the clocks ... the case is the same as with his measuring rod. ... The motion of two equal clocks, one in [the stationary] and the other in [the moving system], will be such that the effective coordinates of the moving parts are, in both systems, the same functions of the effective time. Consequently, if the hand of the [stationary] clock ... returns to its initial position after an interval of time $\theta$, the hand of the [moving] clock ... will do so after an increment equal to $\theta$ of the effective time $t'$. Therefore, a clock in

---

[30] LORENTZ 1952C, 200; CF. LEHTI 2002, 146.

[31] LORENTZ 1952C, 223-224.

> the [moving] system *S* will indicate the process of the effective time, and without his knowing anything about it, [the moving observer's] clocks will go *k* times slower than those of [the stationary observer].

This is central for Lorentz's thinking. A *clock* is a 'mesocosmic' object (Einstein wrote about a watch). Lorentz will show that, because of the laws which govern the microstructure of mesocosmic clocks, they go 'wrong' when moving. Only after a study of Einstein's theory did he realize that mesocosmic clocks show when moving 'real time' as defined by Einstein.

Einstein begins his *Jahrbuch* -article with an exposition of the difficulties of Lorentz's theory concerning 'absolute motion' relative to the ether:

> Astonishingly, it came out that it was necessary only to understand the concept of time sharply enough to overcome the difficulty mentioned. It was only necessary to realize that one can define as 'time' simply an auxiliary quantity introduced by Lorentz, which he called 'local time'.[32]

The attitude of Lorentz has been contrasted to that of Einstein. The contrast (at least to the later Einstein) is maybe not as absolute as has been maintained. Einstein wrote in 1949 something very interesting about clocks:

> First, a remark concerning the theory as it is characterized above. One is struck [by the fact] that the theory (except for the four-dimensional space) introduces two kinds of physical things, i.e., (1) measuring rods and clocks, (2) all other things, e.g., the electromagnetic field, the material point, etc. This, in a certain sense, is inconsistent; strictly speaking measuring rods and clocks would have to be represented as solutions of the basic equations (objects consisting of moving atomic configurations), not, as it were, as theoretically self-sufficient entities. However, the procedure justifies itself because it was clear from the very beginning that the postulates of the theory are not strong enough to deduce from them sufficiently complete equations for physical events sufficiently free from arbitrariness, in order to base upon such a foundation a theory of measuring rods and clocks. ... But one must not legalize the mentioned sin so far as to imagine that intervals are physical entities of a special type, intrinsically different from other physical variables (...).[33]

---

32 EINSTEIN 1989, 435.

33 EINSTEIN 1957, 59.

The definition of clocks and measuring sticks as physical systems consisting of elementary particles and governed by the forces between them was what Lorentz tried to accomplish. He criticized Einstein for not having done that. The difference between the attitudes of Lorentz and Einstein were perhaps not so deep as some have attempted to make them seem.

### 3.3. Henri Poincaré

Another precursor of Einstein was Henri Poincaré:

> In 1904, only one year before Einstein's paper, … the organizers of the St. Louis International Exposition sponsored an International Congress of the Arts and Sciences. … One of the invited was Henri Poincaré, ….[34]

In his lecture, Poincaré explains the problems of 'absolute velocity', principle of relativity, the experiments performed for their elucidation, and the difficulties in interpreting the results. Then he continues:

> The most ingenious idea has been that of local time.
>
> Imagine two observers who wish to adjust their watches by optical signals; they exchange signals, but … they know that the transmission of light is not instantaneous, ….
>
> When the station *B* perceives the signal from the station *A*, its clock should not mark the same hour as that of the station *A* at the moment of sending the signal, but this hour augmented by a constant representing the duration of the transmission. … The watches adjusted in that manner do not mark, therefore, the true time; they mark what one may call the *local time*, so that one of them goes slow on the other. It matters little, since we have no means of perceiving it. …[35]

Poincaré accepts throughout the 'conspiracy theory of nature'; nature prevents the discovery of absolute space and absolute motion, although they *exist*. He accepts *the same physical instant* as an admissible concept (whence 'absolute' simultaneity exists). The *local time* is not the 'correct' time, the principle of relativity means the existence of a 'conspiracy' to the effect that the observer "will have no means of knowing whether he is at rest or in absolute motion". Many draw the conclusion that the theory of relativity was

---

[34] WILLIAMS 1979, 39-40.
[35] POINCARÉ 1979, 41-42.

not yet discovered. That depends on our definition of what the theory of relativity is. Edmund Whittaker[36] comes to the contrary conclusion and holds Poincaré and Lorentz as the discoverers of the theory.

## *4. Einstein's Mesocosmic time: Interpretations and reactions*

### 4.1. The 'semantic' and 'realistic' interpretations

Einstein's friend Philipp Frank has written a biography of Einstein, published in English translation in 1947.[37] Frank gives a statement about what in his opinion the 'Einsteinian reform' was about:

> Many authors have interpreted Einstein's clear and unequivocal statement by the apparently profound but in reality meaningless statement: "Einstein said that sometimes time flows rapidly and sometime slowly". Indeed, to say that time *flows* is a figure of speech that is only partly appropriate to the description of the physical phenomena. … If one differentiates between statements about new physical occurrences and the proposal for a new mode of expression, one can formulate what is exactly meant by claiming the "relativity of time". … Einstein's *relativity of time* is a reform in *semantics*, not in metaphysics.[38]

It remains a problem whether semantic simplicity is still on the side of Einstein when we begin to speak about cosmology. Frank gives a markedly positivistic presentation of the question.

> By people who are interested in the philosophical aspects of relativity theory, the question has been raised: Has science really proved that time is relative and not absolute? The answer can be given quite directly.
>
> … From the theoretical and experimental results of Einstein's principles it becomes evident that the statement that "there is a time distance of 10 seconds between two given events" has no operational meaning. We must always add explicitly a particular system of reference. Therefore Einstein suggested dropping entirely sentences of the form "the time distance is 10 seconds" from the language of physics and of admitting only sentences of the form "the time distance is 10 seconds relative to the system of reference *S*". This suggestion is not a statement about physical facts which can be confirmed or refuted by experiments. It is a suggestion for using a language which is recommended as being well adjusted to our experience about physical facts.

---

36 WHITTAKER 1960.

37 FRANK 1967.

38 FRANK 1967, 63.

> Nobody can be persuaded to accept this recommendation if he does not like to. … Either we concede that we do not know which of the arbitrary systems of reference is the "real" one and leave the discovery of this system to a superior spirit, or we single out some convenient system (e.g. the system of fixed stars) and give to the time distance relative to this particular system the honorary title "real or absolute time distance" while we call the time distance relative to other systems "relative or apparent time distance". Nobody can prove that this way of speaking is "false". … [39]

There remains the question whether a particular reference system of fixed stars is for quite 'objective' and 'scientific' reasons preferable to others. Perhaps it is not, when we are speaking about a time distance of 10 seconds or, still worse, a nanosecond between two given events, but it is if we change the time scale to 100 million years.

Another friend of Einstein, Max Born, published in 1920 an exposition of the theory of relativity; we cite the English translation.[40] Born gives a presentation of the 'realistic' view of Einstein's assumptions, according to which they were not just a semantic matter:

> It might be imagined that Einstein's theory furnishes no new knowledge about things of the physical world but is concerned only with definitions and conventions which are suggested by and in agreement with the facts but which might equally well be replaced by others. … As a matter of fact, if we use sound signals to regulate the clocks, Einstein's kinematics can be applied in its entirety to ships that move through motionless air. … We should have before us a consistent Einsteinian world on a small scale.
>
> But this world would be consistent only so long as we admitted that the units of length and time are to be restricted by no postulates other than the two principles of relativity and the constancy of the velocity of sound or light. Is this the meaning of Einstein's theory?
>
> Certainly not. Rather it is assumed as self-evident that a measuring rod which is brought into one system of reference *S* and then into another *S'* under exactly the same physical conditions would represent the same length in each (provided it is affected in neither system by external forces). … Exactly the same would be postulated for the clocks. …
>
> According to Einstein, … in the real world …, the *same* rod and the *same* clock are used first in the system *S* and then in the system *S'* to fix lengths and times. This is a feature of Einstein's theory by which it rises above the standpoint of a mere convention and asserts definite properties of real bodies. This gives it is fundamental importance for the whole physical view of nature.[41]

---

[39] FRANK 1950, 31-32.

[40] BORN 1962.

[41] BORN 1962, 251-252.

That "Einstein's theory ... asserts definite properties of *real bodies*" is not contrary to the possibility that our speaking about *time* may be a semantic convention. The views of Frank and Born are not contradictory. Frank describes the impression of Einstein's views:

> In the fall of 1912 I first realized that Einstein's theory of the "relativity of time" was about to become a world sensation. At that time, in Zurich, I saw in a Viennese daily newspaper the headline: "The Minute in Danger, a Sensation of Mathematical Science". In the article a professor of physics explained to an amazed public that by means of an unprecedented mathematical trick a physicist named Einstein had succeeded in proving that under certain conditions time itself could contract or expand, .... Some rejoiced that anything so absurd could happen and that traditional science, which is always unpopular with some people, had suffered such a defeat. ... At any rate, it was exiting that something of the sort could happen and that our generation was chosen to witness the overthrow of the foundations of the universe.
>
> How did it happen that something of this kind, in part exciting and in part absurd, was ascribed to Einstein's theory? ... Why did people like to present Einstein's clear deductions about physical experiments in a semi-mystical and incomprehensive language?
>
> The reason is that Einstein not only asserted the existence of previously unknown physical occurrences, but also proposed to describe these new phenomena in a language by which they might be expressed most simply. ...[42]

The nervous reactions of these writers show that the real novelty of Einstein's view was not at all understood. Behind the formulations is the Newtonian concept of time as an existing entity. Einstein did not give new and disturbing properties to this entity; he denied its existence and replaced it with clocks. Had everything been formulated using the concept of a clock, it would not have looked so 'absurd'.

## 4.2. The 'fatalistic' view of the world

In the book *Interpretations and Misinterpretations of Modern Physics* Frank criticized again the views some authors have connected with the theory of relativity:

> One of the most wide-spread metaphysical interpretations of the theory of relativity, ... is that which undertakes to prove the fatalistic conception of the world, that is to say,

[42] Frank 1967, 61-62.

> the view that everything that happens is determined from all eternity, and that there is no development and nothing really new in this world.
>
> ... we may turn at random to any collection of utterances of philosophers about questions of the day, for instance, ... we find the following passage ... by F. Lipsius (...): "The question arises whether, thereby, space has lost its independence of time altogether. ... Minkowski's sentence, "Henceforth, space by itself and time by itself are doomed to fade away into mere shadows and only a kind of union of the two will preserve an independent reality" suggests this view. ... This consideration reintroduces into philosophy the doctrine of the unreality of change and declares all development to be illusion."
>
> ... These metaphysical interpretations of the theory of relativity have their origin in the insufficiently clear formulations which can be found in treatises of physics themselves, ... .[43]

We note that Lipsius takes for granted that the unreality of *time* means the unreality of *change*. Mach early saw the difference. Frank gives an example of how curious interpretations may be found in the texts of physicists:

> As an example I will therefore not quote a philosopher but the well known physicist and astronomer Sir James Jeans who, ... (1935) ... also refers to the theory of relativity as an argument in favour of fatalism. Jeans starts out from the fact that, according to this theory, two events E and F may happen simultaneously for an observer $B_1$, and yet one after the other for a different observer $B_2$. ... Thus one and the same event E, which $B_1$ only expects to occur, may already have happened and be over for the observer $B_2$. This, to cite Jeans literally, is formulated as follows: "It is meaningless to speak of the facts which are apt to come... and it is futile to speak of trying to alter them, because, although they may be yet to come for us, they may already have come for others." And he at once draws conclusions from this for human behavior. He says in fact: "Such a view reduces living beings to automata." And he describes in poetical style how man has changed from a participant into a mere spectator in the world theatre.[44]

We see that the question whether human beings are "participants" or "mere spectators in the world theatre" does not necessarily originate from quantum mechanists but can also rise in quite a different context. According to Newton, time 'exists'. According to the 'absolutistic interpretation of relativity theory', only space-time exists. According to Einstein, only clocks exist. Frank criticizes the meaning of the word 'exist':

---

[43] FRANK 1938, 47.

[44] FRANK 1938, 48-49.

People talk as if the four-dimensional world continuum might "exist" in the same way as a real empirical body exists. They go so far as to say: "the really existing is not the three-dimensional world of bodies but the four-dimensional space-time world".[45]

## *5. Minkowski's Theory of Space and Time*

The mathematician Hermann Minkowski presented a radical rejection of Lorentz's position in a lecture about space and time in 1908.[46] The most important contribution of Minkowski is the concept of *proper time.* Minkowski defines what was later called the 'line-element' *dt* of a "world-line of a substantial point", and writes:

The integral $\int d\tau = \tau$ of this amount, taken along the world-line from any fixed starting-point $P_o$ to the variable end-point P, we call the proper time of the substantial point at P. ...[47]

The physics of moving clocks, or the generalization to all moving systems, whether physical or biological, is based on Minkowski's fundamental discovery of proper time. To Minkowski's address Arnold Sommerfeld wrote a comment in which he has the following to say about proper time and moving clocks:

As Minkowski once remarked to me, the element of proper time *dt* is not a complete differential. Thus, if we connect two world points *O* and *P* by two different world-lines 1 and 2, then

$$\int_1 dt \neq \int_2 dt .$$

If 1 runs parallel to the *t*-axis, so that the transition in the chosen system signifies rest, it is evident that

$$\int_1 dt = t, \quad \int_2 dt < t .$$

On this depends the retardation of the moving clock compared with the clock at rest. The assertion is based, as Einstein has pointed out, on the unprovable assumption that the clock in motion actually indicates its own proper time, i.e. that it always gives the time corresponding to the state of velocity regarded as constant, at any instant.[48]

---

45 FRANK 1938, 52-53.

46 MINKOWSKI 1952; cf. LEHTI 2002, 151-153.

47 MINKOWSKI 1952, 85.

About time dilatation Einstein writes:

> As a generalization we may deduce: All progress of events in a physical system slows down when we give that system a translational motion. The retardation, however, exists only from the point of view of a coordinate system (an observer) not partaking in the motion.[49]

Einstein *deduces* the slowing down of all physical systems in motion. On what is such a deduction based? The generalization mentioned is rather a fundamental physical axiom than a deduction from something else. The paper *Die Relativitäts-Theorie* reproduces a lecture held on 16.1.1911. Here we are told about clocks in motion:

> The affair becomes more bizarre when we think the following performed: We give such a clock a very great velocity (almost like *c*), and let it fly away in uniform motion, and then, after a flight of great distance, give it an impulse in the contrary direction. Then it returns to the place where it originally was thrown away. It comes out that during the entire travel, the hands of this clock have scarcely changed their position at all, although the position of the hands of a clock of the same design, remaining in the place of hurling, has markedly changed. We must add that what holds true for this clock we have introduced as a simple representant for all physical progress of events, holds also for a closed physical system of any other kind. If we for instance would lock a living organism in a box and let it perform the same forth and back motion as the clock above, we would achieve that this organism would return to the original site having changed arbitrarily little, but an organism exactly of the same kind staying at rest in the original site would already long ago have given place to new generations. For the moving organism the long travel time was only an instant, if the motion had approximately the speed of light.[50]

## *6. Einstein about microcosmic time*

### 6.1. Einstein's first reference to 'microcosmic clocks'

In an article of 1907 *Über die Möglichkeit einer neuen Prüfung des Relativitätsprinzips* Einstein refers to an experiment by Johannes Stark in the previous year 1906. The experiment concerned the Doppler-effect of the line-spectrum of channel-rays (*Kanalstrahlen*). Einstein writes:

---

[48] SOMMERFELDT 1952, 94.
[49] EINSTEIN 1995, 545.
[50] EINSTEIN 1993, 436.

In the following I will briefly show that the Principle of Relativity combined with the Principle of constant velocity of light allow the prediction of that phenomenon. ... It follows from those principles that a clock moving with constant velocity, when judged from a 'stationary' system, goes slower than judged by an observer moving with the clock. If, for an observer at rest, $\nu$ denotes the number of tickings of the clock in an unit of time, and $\nu_0$ the corresponding number for the observer traveling with the clock, we have

$$\nu/\nu_0 = (1 - (v/V)^2)^{1/2}$$

or in first approximation

$$(\nu - \nu_0)/\nu_0 = -\tfrac{1}{2}(v/V)^2 .$$

The atomic ion emitting and absorbing Channelrays (*Kanalstrahlen*) of certain frequencies is now to be regarded as a rapidly moving clock, and therefore the relation given above can be used.

We must, however, note that the frequency $\nu_0$ (for the observer moving with the clock) is not known, whence the relation given above is not directly available for experimental control. We must, however, assume that $\nu_0$ is also equal with the frequency which the same ion emits and absorbs when in stationary state .... From that we directly deduce, using the relativity principle, that $\nu_0$ must be equal to the frequency of the radiation emitted or absorbed by a stationary ion.

Therefore also the equation

$$(\nu - \nu_0)/\nu_0 = -\tfrac{1}{2}(v/V)^2$$

gives directly the second order effect searched for.[51]

In his *Jahrbuch*-article, Einstein again refers to Johannes Stark's experiments concerning the channel rays. Einstein reacts as follows:

... we may look at such an ion as a clock having a certain frequency number $\nu_0$. This can be obtained for instance studying light emitted by a similar ion at rest relative to the observer.[52]

When Einstein writes that "we may look at such an ion as a clock" he forms a new hypothesis connecting 'microcosmic time' as defined by the atomic clock with the 'mesocosmic time' defined earlier using a visible clock which has moving hands. In fact, here he does not at all use the time of the coordinate system he had earlier constructed; he requires the assumption that the time parameter regulating the internal processes of a moving microscopic system behaves as the 'time dilatation' demands. He can not *deduce* this with the construction he used when defining time as a reading of

[51] EINSTEIN 1989, 402.
[52] EINSTEIN 1989, 444.

clocks. In a lecture 16.1.1911 Einstein refers again to this 'microcosmic phenomenon' as an immediate consequence of the theory of relativity:

> From the physically important consequences of the Relativity Theory we must still mention the following one. We have already seen that according to that theory, a moving clock runs slower than the same clock when at rest. It might be forever impossible to verify this with a pocket watch, because the velocities possible to give to such one are immeasurable small. Nature, however, offers us objects having the character of clocks which, moreover, can be given rapid motion. These are atoms emitting spectral lines, using electric fields we can give them velocities of several thousand kilometers ⟨per second⟩ (channel rays). The theory lets us expect that the motion of such atoms makes their frequencies appear changed exactly in the same way we can deduce to happen in moving clocks. Although the pertaining experiments meet serious difficulties, we are allowed to hope that in the next decades we arrive either to an important confirmation or a rebuttal of the theory of relativity.[53]

## 6.2. Time and gravitation

Later in the *Jahrbuch*-article Einstein examines the concepts of space and time in a reference system moving with constant acceleration. He gives arguments for the following:

> The principle of constant velocity of light can be used also here for the definition of simultaneity, if we restrict us on very short light paths.[54]

Doing this, we can define a 'local time' s in the system:

> The physical significance of the local time s is, as we immediately realize, the following. If we use that local time s when characterizing events occurring in some singular space-element of Σ, then the laws which those events obey, do not depend on the location, i.e. the coordinates, of the space-element in question, … .[55]

In addition to this local time Einstein also defines a 'coordinate time' $\tau$ using the comparison of the readings of distant clocks with one located at the origin *O* of the coordinates. The relation between the two kinds of time is

---

[53] EINSTEIN 1993, 437.

[54] EINSTEIN 1989, 478.

[55] EINSTEIN 1989, 478.

complicated.[56] Using the equivalence of a gravitational field and the acceleration of the coordinate system, Einstein obtains between the two kinds of time the relation $\sigma = \tau(1 + \Phi/c^2)$, where $\Phi$ is the gravitational potential. Einstein wonders which one of the two 'times' defined should be used in various cases (p. 480). In the *Jahrbuch*-article Einstein writes:

> If in a point *P* where the gravitational potential is $\Phi$, is located a clock showing the local time, then ... it runs $(1 + \Phi/c^2)$ times faster than a similar clock located in the origin of the coordinates. ...
>
> Now there exist clocks in places of various gravitational potential, and their rates can be controlled very accurately: they are the producers of spectral lines. From the preceding we deduce that the light coming from the surface of the Sun, arriving from such a producer, has a wavelength about two parts in a million greater than the light produced by the same elements on the Earth.[57]

This was Einstein's first prediction of an effect which later got the name *gravitational redshift*.

Einstein published an article *Über den Einfluss der Schwerkraft auf die Ausbreitung des Lichtes* in 1911.[58] There he presented two effects following from a combination of the theory of gravitation and the Principle of relativity. They were the bending of light and the shifting of spectral lines in gravitational fields. Einstein concludes that when a ray of light is emitted from a place with a definite gravitational potential, and possessing at its emission certain frequency, it will at its arrival in a location with a different gravitational field possess a different frequency, measured by an identical clock. This result gives rise to considerations concerning the behavior of clocks in a gravitational field. It seems to assert an absurdity. If there is a constant transmission of light from one place to another, how can any other number of periods per second arrive to the latter than is emitted in the former? The answer is that we cannot regard the frequencies simply as the number of periods per second since we have not yet determined the nature of the time in all reference systems. One frequency denotes the number of periods with reference to one time-unit of a clock, and the other denotes the number of periods per second with reference to the identical clock in another location. Nothing compels us to assume that the clocks in different gravi-

---

[56] EINSTEIN 1989, 478.

[57] EINSTEIN 1989, 480-481.

[58] EINSTEIN 1952c

tation potentials must be regarded as going at the same rate. We must define the times in such a way that the number of wave crests and troughs is independent of the absolute value of time. This leads to the change of the rate of the clock and, hence, to that of the 'time' in each gravitational field.

## *7. Megacosmic time*

### 7.1. Stationary and expanding universes

Einstein's cosmological paper of the year 1917 dealt with the problems inherent in trying to extend his gravitational equations to the entire universe:

> Nevertheless, there is still a possible way out, without resigning .... For if it were possible to regard the universe as a continuum which is *finite* (*closed*) *with respect to its spatial dimensions*, we should have no need at all of any such boundary conditions. ...[59]

Such a universe Einstein describes.[60] He has taken for granted that the universe has *spatial dimensions*, and, as an implication, a *time dimension*, and these are independent of any observer and his motion. That assumption has since then been made in practically all cosmic models[61] .

We told earlier how Philipp Frank criticized James Jeans' metaphysical speculations. Then he gives interesting information of how Jeans thought modern cosmology saves us from determinism.

> ... According to Jeans, however, it is a fortunate fact for a mankind discouraged by fatalism, that the progress of physics, or, more precisely, astronomy, has, after all, proved that the hypothesis of the eternal existence of the entire space-time world was mistaken. "If the theory of relativity", Jeans continues, "was to enlarged so as to cover the facts of astronomy, then the symmetry between space and time which had hitherto prevailed, must be discarded. Thus time regained a real objective existence, although only on the astronomical scale, and with reference to astronomical phenomena". ... Of this asymmetry Jeans goes on to say: "This gives us every justification for reverting to our old beliefs that past, present, and future have real objective meanings, and are not mere hallucinations of our individual minds – in brief we are free to believe that time is real."[62]

---

[59] EINSTEIN 1952f, 183.
[60] EINSTEIN 1952f, 183-186.
[61] Cf. LEHTI 2002, 156-161.
[62] FRANK 1938, 55.

Frank goes on to argue that not even the symmetry of the line-element could be used to show the eternal existence of the future, but that point does not diminish the interest of Jeans' last arguments. They show that already since Einstein's cosmological article from 1917, together with de Sitter's contemporaneous works, it was clear that in astronomy an objective and universal division of four-dimensional space-time into space and time was well known. How are we to interpret the assertion that the objectivity of one single "time" must be restricted to only "the astronomical scale, and with reference to astronomical phenomena"? How can more or less local physical phenomena and the global astronomical phenomena behave as if they were phenomena of two different worlds?

The cosmologist McVittie gives a class of solutions of the geodesic equations in general relativity theory and continues:

> ... The observer, permanently located at the origin of spatial coordinates, also has a special world-line. ... Matters may then be so arranged that the proper-time *s* along every special world-line has the same zero value at the initial instant. Thus a common time for all special world-lines is introduced ... . [63]

McVittie has *de facto* defined an 'absolute time' as *cosmic time*; i.e.the common time of all coordinate systems immobile in the expanding space. He continues:

> ... The travel-time of radiation received from any galaxy is believed to be immensely long compared with the entire history of astronomy. Hence it is sufficient to assume that the observer makes all his observations at a single instant $t_o$.[64]

This renders the use of *cosmic time* impracticable as a time parameter for any processes upon the Earth as studied by physicists. This isolation of the immense 'megacosmic' time scale of cosmology from the 'human' or 'mesocosmic' as well as from the 'microcosmic' time scale used by physicists makes it in practice possible to handle the former in terms of 'absolute time', and the latter in terms of relativistic space-time. Here in principle is a discrepancy. This thinking leads to the assumption that *all* physical 'mesocosmic' or microcosmic theories having an inbuilt parameter of 'time' are idealizations. Their mathematical properties, such as the 'principle of rela-

[63] MCVITTIE 1965, 145.
[64] MCVITTIE 1965, 150.

tivity', 'time symmetry', etc, should not be taken as properties of the real world. They are properties of a mathematical model and have restricted instrumental value.

Two astronomers Yurij Baryshev and Pekka Teerikorpi write in 2002 about time:

> The common cosmic time for observers rushing away from each other also looks paradoxical, because in laboratory physics moving observers have different observed times and it is a chief tenet of relativity theory that there is no universal absolute time. But for the universe as a whole there is a kind of preferred time. ...
>
> ... It seems that in spite of the principle of relativity, which is in excellent agreement with everything that the physicist observes in his laboratory, the universe has been so structured that there is a space relative to which one may measure our velocity as about 400 km/sec. ... [65]

The authors, however, warn the readers:

> ... And, of course, the universe is no object of classical physics, even if the circle seems to be closed, when one encounters something like "absolute time", "absolute space" and "velocities greater than the speed of light", but now on a higher level of reality.[66]

We have not returned to classical physics, but the epistemological arguments with which philosophers (and also many physicists) have battered the concepts of 'absolute' space and time have proved problematic. The language of 'higher levels of reality' is questionable, but the 'megacosmic clock' is a concept quite different from Newton's absolute time. Maybe the most crucial change is the assumption that 'clock' is the primary concept and 'time' the secondary concept.

## 7.2. Temperature of microwave radiation as a marker of cosmic time

Cosmic background radiation is measured by the "Planck spectrum". In 1980 the corresponding temperature was measured at 2.96° K. (The value has of course been improved later.) Steven Weinberg writes in 1983:

---

[65] BARYSHEV & TEERIKORPI 2002, 211.

[66] BARYSHEV & TEERIKORPI 2002, 212.

> Oddly enough, the radiation background *was* measured at shorter wavelengths, long before any of the astronomical work discussed so far in this chapter, ... In the constellation Ophiucus ('the serpent bearer') there is a cloud of interstellar gas which happens to lie between the earth and a hot but otherwise unremarkable star, ζ Oph. The spectrum of ζ Oph. is crossed with a number of unusual dark bands, indicating that the intervening gas is absorbing light at a set of sharp wavelengths. ... Thus, observing the wavelengths where the dark bands occur, it is possible to infer something about the nature of these molecules, and of the states in which they are found.
>
> One of the absorption lines in the spectrum of ζ Oph … [indicates] the presence in the interstellar cloud of a molecule, cyanogens (CN), … In 1941 it was found … that this absorption line is actually split, consisting of three components …. McKellar was able to estimate that the cyanogens was being exposed to some sort of perturbation with an effective temperature of about 2.3° K, which could lift the cyanogens molecule into the rotating state.
>
> … After the discovery of a 3° K cosmic radiation background in 1965, it was realized (…) that this was just the perturbation that had been observed in 1941 to be producing the rotation of the cyanogens molecules in the Ophiucus clouds. … [67]

These early observations gave values for cosmic microwave radiation temperature which are crude when compared to later results. Now we can go from the assumption that the radiation is black-body radiation, and *at least in principle* observations as described above allow us to calculate the temperature of the radiation at the time it had the observed effects. In principle the temperature must have been higher than at present, and if the phenomena observed are distant enough, the difference in temperature might be possible to determine. The fall of the temperature is a measure of the expansion of the universe and, therefore, also a measure of the age of the universe, i.e, *time* as measured by the 'cosmic clock', the universe itself. In this way we may measure distant time without need of 'spreading' it; it is already there. Since the observation described by Weinberg much has happened. John Bahcall writes:

> The idea of using atomic fine-structure lines to test the Big Bang cosmology was first suggested in 1968, only three years after the discovery of the CMB. Over the past three decades, a series of ever more precise measurements, or upper-limit determinations of the CMB temperature in distant clouds were made using fine-structure lines. Generations of sensitive telescopes and spectroscopic instruments had to be developed before the last robust test was possible. … [68]

---

[67] WEINBERG 1983, 73-74.

[68] BAHCALL 2000, 917.

The 'robust test' was made in 2000 by the Indian astronomer Raghunathan Srianand and his coworkers at the Paranal Observatory in South America. Their raport begins:

> The existence of the cosmic microwave background radiation is a fundamental prediction of the hot Big Bang cosmology, and its temperature should increase with increasing redshift. At the present time (...), the temperature has been determined with high precision to be … 2.726 ± 0.010 K. … The background temperature can be determined using measurements of the relative positions of atomic fine-structure levels, which are excited by the background radiation. … Here we report the detection of absorption lines from the first and second fine-structure levels of neutral carbon atoms in an isolated cloud of gas at [redshift] 2.3371. … These … enable us to determine that the background radiation was indeed warmer in the past: we find that [the temperature] is between 6.0 and 14 K. This is in accord with the temperature 9.1 K predicted by hot Big Bang cosmology.[69]

These results mean that simultaneity in the cosmos can be defined in a quite absolute and in principle 'operational' way as equality of the temperature of cosmic microwave radiation. Some function of the temperature itself could then be defined as *time*. It should, of course, be a monotonously decreasing function (for the time to increase in the direction to the future). It is practically irrelevant which function we take, because in the entire time span of human civilization only the differential of that function matters. Therefore, let us take the simplest function $1/x$ and give a quite operational definition of (cosmic) time: *Time is the inverse value of the Kelvin-temperature of microwave cosmic radiation.* This definition side-steps all difficulties in transportation of clocks or spreading time into space.

## *Bibliography*

BAHCALL, JOHN
2000 "The Big Bang is bang on" in *Nature* vol 408.21, 28 December. 916-917.

BARYSHEV, YURIJ & TEERIKORPI, PEKKA
2002 *Discovery of Cosmic Fractals*. Singapore: World Scientific.

[69] SRIANAND, PETITJEAN & LEDOUX 2000, 931.

BERKELEY, GEORGE
1998 *A Treatise Concerning the Principles of Human Knowledge*. Ed. Jonathan Dancy. Oxford: Oxford University Press.

BONDI, HERMANN & WESTON-SMITH, MIRANDA (eds.)
1998 *The Universe Unfolding*. Oxford: Oxford University Press.

BORN, MAX
1962 *Einstein's Theory of Relativity*. New York: Dover Publications.

EINSTEIN, ALBERT
1952a "On the Electrodynamics of Moving Bodies" in Lorentz, Einstein, Minkowski & Weyl 1952, 35-65.
1952c "On the Influence of Gravitation on the Propagation of Light" in Lorentz, Einstein, Minkowski & Weyl 1952, 97-108.
1952f "Cosmological Considerations on the General Theory of Relativity" in Lorentz, Einstein, Minkowski & Weyl 1952, 175-188.
1957 "Autobiographical Notes / Autobiographisches" in Schilpp 1957, 1-95.
1982 *Ideas and Opinions*. New York: Three Rivers Press.
1989 *The Collected Papers of Albert Einstein*. Volume 2. The Swiss Years: Writings, 1900-1909. Ed. John Stachel. Princeton: Princeton University Press.
1993 *The Collected Papers of Albert Einstein*. Volume 3. The Swiss Years: Writings, 1909-1911 Eds. Martin J. Klein, A. J. Kox, Jürgen Renn and Robert Schulmann. Princeton: Princeton University Press.
1995 *The Collected Papers of Albert Einstein*. Volume 4. The Swiss Years: Writings, 1912-1914. Eds. Martin J. Klein, A. J. Kox, Jürgen Renn, and Robert Schulmann. Princeton: Princeton University Press.
2001 *Relativity. The Special and the General Theory*. Transl. Robert W. Lawson. New York: Routledge.

FRANK, PHILIPP
1938 *Interpretations and Misinterpretations of Modern Physics*. Actualites scientifiques et industrielles 587. Paris: Hermann & Cie, Éditeurs.
1950 *Foundations of Physics*. International Encyclopedia of Unified Science Volume I, Number 7. Chicago: University of Chicago Press.
1967 *Einstein; his Life and Times*. New York: Alfred A. Knopf.

LEHTI, RAIMO
2002 "The Concept of Space in Relativity Theory and Cosmology" in Martikainen (ed.) 2002, 143-153.

LORENTZ, HENDRIK ANTOON
1952b "Electromagnetic Phenomena in a System Moving with any Velocity less than that of Light" in Lorentz, Einstein, Minkowski & Weyl 1952, 9-34.
1952c *The Theory of Electrons*. New York: Dover.

LORENTZ, HENDRIK ANTOON & EINSTEIN, ALBERT & MINKOWSKI, HERMANN & WEYL, HERMANN
1952 *The Principle of Relativity, a collection of original memoirs on the special and general theory of relativity*. Transl. W. Perrett and G. B. Jeffery. New York: Dover.

MACH, ERNST
1960 *The Science of Mechanics*. La Salle, Ill.: Open Court.

MARTIKAINEN, EEVA (ed.)
2002 *Infinity, Causality and Determinism. Cosmological Enterprises and their Preconditions.* Frankfurt am Main: Peter Lang.

MCVITTIE, G. C.
1965 *General Relativity and Cosmology*. London: Chapman & Hall.

MINKOWSKI, HERMANN
1952 "Space and Time" in Lorentz, Einstein, Minkowski & Weyl 1952, 73-91.

NEWTON, ISAAC
1952 *Opticks*. New York: Dover.
1966-1973 *Sir Isaac Newton's Mathematical Principles of Natural Philosophy and his System of the World I-II.* Transl. Andrew Motte, revised by Florian Cajori. Berkeley: University of California Press.

POINCARÉ, JULES HENRI
1979 "The Principles of Mathematical Physics" in Williams 1979, 39-49.

SCHILPP, PAUL ARTHUR (ed.)
1957 *Albert Einstein: Philosopher - Scientist.* New York: Tudor Publ. Co.

SCHLICK, MORITZ
1963 *Space and Time in Contemporary Physics.* New York: Dover.

SOMMERFELD, ARNOLD
1952 "Notes" to Minkowski 1952 in Lorentz, Einstein, Minkowski & Weyl 1952, 92-96.

SRIANAND R. & PETITJEAN, P. & LEDOUX, C.
2000 "The cosmic microwave background radiation temperature at a redshift of 2.34" in *Nature* vol. 408. 21/28 December. 931-935.

TORRETTI, ROBERTO
1996 *Relativity and Geometry*. New York: Dover.

WEINBERG, STEVEN
1983 *The First Three Minutes. A modern view of the origin of the universe*. London: Fontana.

WHITTAKER, EDMUND
1960 *A History of the Theories of Aether and Electricity I-II*. New York: Harper & Brothers.

WILLIAMS, LESLIE PEARCE (ed.)
1979 *Relativity Theory: Its Origins and Impact on Modern Thought*. Huntington, N.Y.: Robert E. Krieger Publ. Co.

# Time: Physical and Eternal

ATSO A. EERIKÄINEN

One cannot progress very far in answering the questions concerning reality without considering the riddle of time. Time may be the most enigmatic of all metaphysical and physical problems, and it must be resolved in order to understand reality.

It seems that time is a *fundamental variable* in reality. There is no law of physics which does not require the concept of time for its formulation, but needless to say, there is no unanimously accepted definition of time. The humble purpose of this paper concerning *two* variables of time is to provide a step toward a meaningful definition of time.

## *Time as metaphysical problem*

Two opposite views of time have been clashing from the dawn of Western thinking. Heraclitus (ca.540-480 B.C.) thought that the basic feature of reality is "becoming," that is, time is a "flux" or perpetual change. As such, time is something *physical*: measurable, countable, and computable, that is, *a posteriori*. Parmenides (ca. 515-445 B.C.), in turn, thought there is no change in the universe, because the permanent forms of reality are motionless and mathematical. The universe is timeless and hence, it is something *eternal*: immeasurable, uncountable, and incomputable, that is, *a priori*.

Western metaphysicians have mainly argued in favor of Heraclitus. There have been three paradigmatic topics concerning time in philosophical enquiry: change, causation, and modality. [1]

[1] See LE POIDEVIN & MACBEATH (eds.) 1993, next abbreviated: *PT* and EERIKÄINEN 2003, next abbreviated: *TDT*.

*Time as change.* It is commonly accepted that only time, not space, is the variable of change. The genuine change involves temporal variation in the properties of things, when time "passes." Time is the changing *process*, where the future is becoming the present, and the present is changing into the past. It is just the change that goes on in the event while it is occurring. A presentness of an event is its happening, as opposed to its having happened or being about to happen. The present is a factual point in the flux of time or a number of motions, which separates and unifies the past and the future, or earlier and later, as Aristotle (384-322 BC) thought. From the *ontological* difference between the past and the future, follows that the past expands in the flux of time: more facts are added to the totality of facts.

J.M.E. McTaggart showed in his famous paper concerning the unreality of time the movement of time consists in the fact that later and later terms pass into the present, or that the present passes to later and later terms. In other words, the so-called "B series" of time is sliding along a fixed "A series" of time, or A series of time is sliding along a fixed B series of time. In the first case, time presents itself as a moment from the future to the past. In the second case, time presents itself as a moment from earlier to later. The events seem to come out of the future, while we ourselves move towards the future. B series of time runs *backwards*, whereas A series of time runs *forwards*, that is, the future has been, the present is, and the past will be, and *vice versa.*[2]

*Time as causal relation.* The main point in the causal theories of time is that time, as a temporal becoming, is asymmetric, deterministic, and continuous. If A causes or is among the causes of B, then B does not cause or is not among the causes of A, and A is sufficient cause for effect B. The *asymmetry* of time entails that time has a direction because causation has a direction. Asymmetric time is the variable of causality. The direction of asymmetric time means also that the so-called "time travel" is impossible, as already Hume (1711-1776) proposed. Especially, if the present is ontologically privileged or time is unreal, there is nothing for the time-traveler to visit.

Yet within physical time, we cannot affect the past, because it is *determined.* Any statement about the past has a determined truth-value. The future, instead, is not yet determined. There is no present fact about

[2] MCTAGGART 1908: 457-474.

whether A will exist tomorrow, so anything we do or happens now can make a difference for the future. If there is no ontological difference between the present and the future, then future-tense statements must have a determined truth-value. Hence, the rejection of temporal becoming means a rejection of indeterminism, that is, free will.

The flux of time seems to be *continuous*. The changing things in time exist continuously. Kant (1724-1804) proposed in his principle of contradiction that a thing cannot be itself and something else simultaneously. If one state exists, the other cannot exist simultaneously. One cannot be both existing and non-existing, or neither existing nor non-existing simultaneously. Reality is a continuity of the temporal states.

*Time as modal relation.* Finally, time can be described in modal terms. "Necessarily p" means that p is true in all possible worlds, and "possibly p" means that p is true at least in some world. The future is just a set of possible worlds, so the flux of time is the passage from the possible to the necessary. However, if causality is the mechanism by which we have knowledge of temporal relations, then there is some epistemological motivation for a non-modal causal theory of time. One might ask why it is necessary to base causal relations on the possibility between events and not the *actual* temporal relations between them. It is only because time somehow exists between two events that there is the possibility of change between those events. Yet, there are some topological difficulties between time and modality. Is the structure of time *necessary* or *contingent*? Aristotle argued in favor of necessity, but Hume in favor of contingency. Hume also thought that time is merely composed by discrete instants, and hence, it cannot be continuous.

The followers of Parmenides have argued in favor of unchanging, unreal, and/or non-existing time. For instance, McTaggart thought that impressions of temporality are mere human inventions. Because time cannot change in time, there is no time existing independently. Temporal relationists believe that time is only the *measure* of change or motion. Time does not change, and its existence depends on motion, as Aristotle thought. As such, time is a *scalar* quantity. Leibniz (1646-1716), in turn, thought that temporal relations can be logically analyzed as *timeless* properties of individual monads. Time is *unreal* but not empty. It is regarded as a possibility of the temporal location. Kant, in turn, located time into the mind as a *pure form a priori* of sensible intuition or reason. As such, time has no physical status. Whereas Euclidean space is the formal *a priori* condition of outer experiences, time is the formal *a priori* condition of appearances. If this subjective condition of sensibility were removed, time would be nothing at all. Thus, space and time are given *a priori*, whereas everything that is given in them is *a posteriori*.

## *Time as physical problem*

*In classical mechanics* (CM) of Newton (1642-1724), time is *absolute*, true, and mathematical, which of itself and from its own nature, flows without relation to anything external. Material bodies move through Euclidean space along predictable paths, subject to forces that accelerate them in accordance with strict mathematical laws. The universe is a gigantic clock-like mechanism, predictable in every detail by universal and absolute time. Time is simply there, and nothing can affect it. Newtonian time is absolute *calculus*: the precise and continuous succession of the present moments.

Newton's conception of absolute time was rejected 150 years later by Einstein's (1879-1955) flexible time. Time became Riemann's non-Euclidean "metric" space or at least an inseparable part of it. Euclidean and "phenomenalistic" special relativity (SR) needs observers and their time depending on how they are moving, whereas non-Euclidean and "realistic" GR does not need any observers. Strictly speaking, Einstein's general relativity (GR) does not describe time but gravity, which has some important implications concerning time. On the one hand, gravity is not a mysterious mechanical force operating at a distance but a warping of space-time by the mass and energy on it. On the other hand, it is an acceleration that depends on the curvature of space-time. So to say, mass tells space-time how to curve, and space-time tells mass how to move. [3] In fact, Einstein's space-time as a Rie-

[3] According to Riemann's (1826-1866) concept of "metric space," it is possible to think of n-dimensional non-Euclidean spaces, and that every space includes an infinite number of internally different spaces. These differ from each other only in curvature. The curvature of two-dimensional Euclidean space is 0, and therefore the parallel-axiom is valid in it. In Riemannian spaces, the curvature is constant and positive. This kind of space is "elliptic;" for example, the surface of a ball, in which parallel lines cannot be drawn through a point outside a given line. All lines always cross each other, and the sum of angles in triangle is always bigger than 180°. Instead, in spaces of Lobatchevski and Bolyai, the curvature is constant, but negative. This kind of space is "hyperbolic," in which more than one parallel line can be drawn through a point

mannian field is simply there, like Newtonian absolute time, but as relative, because it depends on motion of mass and energy on it. Because, it seems that the velocity of gravitons is equal with the velocity of light, as Einstein predicted, both elastic and especially inelastic interaction of gravitons and mass particles causes the bending effect on light passing through space near a large and massive body. So, it would be that just gravity or both the gravitational pushing and the electromagnetic pulling is the field of space-time. [4]

Einstein's GR imply that the universe and physical time must have a beginning, and they may have an end. Causality determines that physical time must have an irreversible direction. The laws of thermodynamics verify the law of causality. The second law of thermodynamics determines that every closed system tends toward a state of total disorder. Entropy measures the degree of disorder, which always increases with time. If there is no beginning of the universe, it would be in a state of complete disorder by now, and everything would be at the same temperature. [5]

GR implies that physical time is also warping. Because a gravitational field is associated with warping time, it tells how much time is dilated at each point in space. An infinite time warp could occur in the so-called "black hole," from where light would not be able to escape because of the infinite density of that space-time singularity. The event horizon of the space-time singularity is the boundary to physical time itself. It is possible

---

outside a given line, and the sum of angles in a triangle is always smaller than 180°. All these spaces are mathematically equally true. The concept of the three-dimensional non-Euclidean space nullified the mathematical twilight, which had been characteristic for all attempts to solve the problem of space in physics. See Eerikäinen TDT, 52-57. – According to Gary Vezzoli, however, the reason why Riemannian space is more suitable to understanding nature and the universe, and analyzing it, is not because of any inherent improvement over Euclidean or Cartesian space, but because of the nature of gravity and gravitons which "force" mass to move through space in a curved, non-straight-line fashion. Such does not imply that space is curvilinear or warped, nor does it imply that space-time is so geometrized. See http://arxiv.org/abs/physics/0104018.

[4] See Davies, 1996, 16-17, 29-33, 40-44, 51-55, 279-283, Feynman 1997, 93-109, Hawking & Penrose 1996. -Based on the work at the University of Missouri and at the National Astronomical Laboratory in Charlottesville, studying the eclipsing of the deep space quasar Jo842+1835 during the period 7-9 September 2002, particle physicist Gary Vezzoli informed me he feels confident that *the speed of gravity is the same as the speed of light*, as Einstein predicted this decades ago.

[5] Hawking, "The Beginning of Time" in: *http://www.hawking.org.uk/lectures/lindex.html.*

that a singularity, the so-called "Big bang," was once the spontaneous origin of physical time, and the so-called "black hole" singularities and/or "Big crunch" might be the end of physical time. [6]

Instead, in *Quantum mechanics* (QM), time is not the Einsteinian relativistic one but it is Newtonian absolute time, and its geometry is Euclidean. QM describes phenomena according to a special time coordinate, that is, the state f (t) of a system follows Schrödinger's equation:

$df/dt\,(t) + iHf\,(t) = 0, f\,(0) = \text{g}$, where H is Hamiltonian, g the initial state, and t is time of the system.

At the quantum level of reality, there is, however, a basic limit that introduces an irreducible fuzziness to the notions of speed, rate, and time: Werner Heisenberg's (1901-1976) *uncertainty principle.* The size of the wave function at a point gives the probability that the particle will be found at that point, and the rate at which the wave function changes from point to point gives the probability of different velocities. One can have a wave function that is sharply peaked at a point. This corresponds to a state in which there is a little uncertainty in the position of the particle. However, the wave function varies rapidly. It means that there is a lot of uncertainty in the velocity. Similarly, a long chain of waves has a large uncertainty in position, but a small uncertainty in velocity. One can have a well defined position or a well defined velocity but not both at the same time. This would seem to make complete determinism impossible. If one cannot accurately define both the positions and the velocities of particles at one time, how can one predict what they will be in the future? Even if time is absolute Newtonian clock-time in QM, there is no absolute clock in QM, because all physical clocks are subject to quantum uncertainty. Hence, also time itself may be subject to quantum effects, if Heisenberg's principle is valid. [7]

[6] DAVIES, 1996, 33-35, 106-125, 131-132.

[7] DAVIES, 1996, 90-92. Einstein did not believe that there is any uncertainty in the universe. According to Stanley L. Jaki, 1999, 102-105, the uncertainty principle is an obscure notion. It leads to fuzzy conclusions. According to the uncertainty principle, (1) *causality* does not hold in QM. However, the so-called "Big bang" was a quantum mechanical singularity. Hence, there is no causality in the universe. (2) Reality is *discontinuous.* An interaction that cannot be measured exactly, cannot take place exactly. But because the "measurement" is an operation and the "taking place" is an ontological occurrence, the conclusion is an elementary mistake in logic. (3) The uncertainty principle can be written: $\Delta mc^2 > \hbar/\Delta t$, where $m$ is mass, $c$ the constant

## *What is now?*

Einstein was seriously worried about the question: "What is now"? He concluded that the "now" has no physical status, and hence, it was a metaphysical question that lies beyond scientific physics. [8] Instead, Sir Arthur Stanley Eddington (1882-1944) thought that our impression of "becoming" is so powerful and central to our experience that it must correspond to something in the objective world. He thought: "If I grasp the notion of existence because I myself exist, I grasp the notion of becoming because *I myself become*. It is the innermost Ego of all which is and becomes. It seems that we experience time in two distinct ways: *externally* through the senses and *internally* within the mind." [9]

Although Plato (c. 428-347 BC) was obviously the first discoverer of "self" (soul, mind) in Western thinking[10], it was Aristotle who first put the mind into the center of changing reality. For him time was a measure of change, and as such, a number of motions in connection with earlier and later. There is no time without motion, and there is no present without the mind who realizes it.[11] In this sense, Einstein's SR is only a novel variation of Aristotle's metaphysics. But what is the now? Could it be (1) something *physical*: changing, measurable, countable, and computable; or (2) something *eternal*: unchanging, immeasurable, uncountable, and incomputable; or (3) something *physical and eternal* simultaneously? There are some novel theories based on (1), such as "Local time" of Hitoshi Kitada (LT) and "Rhythm Based Time" (RBT) of Peter Beamish, and theories, based on (2):

---

velocity of light, $\hbar$ is Planck constant, and $t$ is time. Because $c$ and $\hbar$ are constants, the factor in evaluating the total uncertainty is only $\Delta m$. If there is no causality in QM, the causality defect becomes a true mass defect in the interaction. If we want to ascertain time $t$ of a particle emission from an atom, we cannot avoid concluding that in the emission, we have on hand an amount of mass which is missing without a cause.

[8] Einstein confessed to Rudolf Carnap (1891-1970) that "What is now worries me seriously", that "there is something essential about the now," but what ever it was, it is "just outside the realm of science." See SCHILPP, 1963, 37.

[9] EDDINGTON, 1929, 97.

[10] In Eastern thinking, it was Gautama Buddha (563?-483? BC) who discovered the innermost ego, and since his time, Eastern philosophy and practice have been focused on the mind and introspection. Especially in metaphysics of Japanese Zen-Buddhism, the mind is seen as "absolutely inconsistent self-identity." See f. ex. JAMES W. HEISING'S anthology *Philosophers of Nothingness. An Essay on the Kyoto School* (Honolulu: University of Hawaii Press, 2001).

[11] See ARISTOTLE, *Physics*, IV: 10-14.

for example, time as a pure form a priori of intuition in Kant's transcendental idealism and a Parmenidian theory of immeasurable time-capsules by Julian Barbour, [12] and at least a prominent theory based on (3): the relative now of secondary becoming or *process* (G) and the absolute now of primary becoming or *transition* (W) of Karl Heim's dimensional theory (HDT) proposed a half century ago.

## *Local time (*LT*)*

According to a Japanese mathematician and metaphysician, Hitoshi Kitada at the University of Tokyo, as far as we are able to know, the microscopic phenomena follow precisely QM. Because the universe is assumed to have begun from an infinitesimally small point within the so-called singularity of "Big bang," we have to apply QM to this initial singularity. However, because the macroscopic universe generated by the singularity follows GR in classical approximation, we need the *relativistic* QM. How it is possible to construct relativistic QM, that is, a quantum theory of gravity has been the main puzzle of physics since Einstein.

According to Kitada, it seems that there are two problems behind the unification of GR and QM: (1) the conflict between their different kinds of geometries, and (2) the conception of time. In order to resolve the geometrical conflict and to define time properly, Kitada introduces his theory of "local time" (*LT*). [13]

As was noticed above, QM is constructed on Euclidean geometry. Therefore, there is a special time coordinate defined without any ambiguity, according to which physical phenomena must be described. Instead, because GR is constructed on non-Euclidean Riemannian geometry, there is a large arbitrariness of choosing space-time coordinates in GR. Because the physical laws are covariant under the transformations between coordinates, any one of them is equivalent in describing physical phenomena. These two notions are obviously incompatible. The resolution of the problem is attempted by using two opposite methods in quantization.

---

[12] In this paper, we do not analyze neither Kant's nor Barbour's conception of time. See a detailed analysis of Kant's ideas and their flaws in *TDT*, 23-32, and 105-111, and Barbour's ideas and their flaws in *TDT*, 138.

[13] See KITADA 2002. Time XI (http://www.kitada.com/), and his other papers about time.

The so-called "canonical quantization" is an attempt to preserve the framework of QM by choosing *beforehand* a special time coordinate for the quantization, but as a result, the covariance is lost. Conversely, the so-called "non-canonical quantization," starts without any special time coordinates. QM is generalized in order to make it covariant with respect to the change of coordinates by defining *afterwards* a special time coordinate in order to recover the quantum mechanical description. In both of these two methods of quantization, it is presupposed that a special time coordinate exists independently in order to make the description of quantum mechanical phenomena effective. Thus, we are back to the question: "What is the definition of time?"

The local systems of Kitada's theory are inner systems equipped with Euclidean geometry and QM associated to each point on the Riemannian manifold of GR. Thus, Kitada adopts Euclidean geometry for QM as an *inner* space and Riemannian geometry for GR as an *outer* space. Time recaptures its meaning as a "local time" inside each local system, defined as a measure of quantum mechanical motion inside the local system. In other words, time is defined as a notion associated with a system that can accommodate clocks, which have actual sizes. [14]

All attempts to unify GR and QM have failed, not only for reasons of their different kinds of geometries, but because time is presupposed to *exist* as an independent or external entity separated from the minds reading it on a clock. Time does not appear without measuring it by a clock. Time is a movement of the hands of the clock, and it does not have an *a priori* existence that measures motion. Just the motion of hands of an analogue clock or just the change of figures of a digital clock measures time. The definition of "local time" is based on this observation, and it is defined as a local notion effective only in each local system. As such, *time is equivalent to the quantum mechanical evolution of a local system.* Since clocks are proper to each local system, and local systems are mutually *independent* concerning the association among the coordinates of these systems, one can impose relativistic change of coordinates among them. The change of coordinates gives a relationship between those local systems which yields relativistic quantum mechanical Hamiltonian, explaining the actual observations. Thus, LT is created by the minds of the observers, and does not demand continuous and external "global time" effective throughout the universe.

---

[14] See KITADA 1999.

LT arises from the *stationary* nature of the universe.[15] The "stationary" of the total universe, that is\Φ, means mathematically an *eigenstate* of a total Hamiltonian H. This, in turn, means that the universe \Φ as the *existence* itself is *timeless*, and as such it cannot be explained in terms of duration of physical time.

Axiom 1. The universe is of *infinite* nature, and it is *eternal*. In other words, the wave function Φ of the universe satisfies with respect to the total Hamiltonian.

$Htotal\,\Phi = \Delta\Phi$, for some non-positive real number $\Delta \leq 0$.

In fact, the eigenstate does not contain any reference to time, as can be seen from its definition. However, this definition does not argue that the entire universe \Φ is any frozen instant, which sempiternally lasts forever without a beginning and an end. Instead, the universe \Φ as the totality of *existence* has the infinite freedom inside itself as an internal motion of finite and local systems. [16]

Axiom 2. The local system is of *finite* nature, having its own origins of position $x$ and momentum $p$, independent of the other's origins and inside worlds. However, the inside world of a finite local system is of *infinite* nature.

The local system oscillates in itself, because the stationary nature of the entire universe assures the oscillation of the locality.

Axiom 3. The oscillating nature of the local system caused by the stationary nature of the universe is expressed by $exp\ (-2\pi itH/h)$, which is known in QM as the evolution operator of the local system. It is the *local clock* of the system, and so, $t$ is *local time* of the system. [17]

---

[15] The quantum mechanical concept "stationary" does not mean motionless, as usually is understood. For example, you can say a hydrogen atom is stationary when it is stable and forms an atom. But inside it, electrons are moving around the nucleus.

[16] See KITADA & FLETCHER, 1996, 38-45. Kitada and Fletcher based on Spinoza's idea of eternal substance (*natura naturans*), which is "exhaustive potentiality exhaustively actualizing itself."

[17] KITADA proves the equation of a local Hamiltonian by using a Fourier transform of a function. Sine and cosine components are summarized to the exponential function $exp\ (-ixp)$, which

The infinite inside world of our local systems is quantum mechanical, and it obeys Schrödinger's equation. Instead, the finite outside world of our local systems obeys the general principle of relativity and the principle of equivalence.

> Axiom 4. General Principle of Relativity: physical worlds or physical laws are the same for all local observers.

> Axiom 5. Principle of Equivalence: for any gravitational force, we can choose a coordinate system as a function of time *t*, where the effect of gravitation vanishes.

Axioms 4 and 5 concern the distortion of our view confronting the finite outside world, whereas the axioms 1-3 concern the infinite inside world, which is independently conceived as its own. The oscillation of the local systems is due to the intrinsic "internal" cause, whereas the distortion of our view toward the outside is due to observational "external" cause. The internal and external aspects are mutually independent from each other, because the internal coordinate system of a local system is a relative one within the local system and it does not have any relationship with the external coordinates. Therefore, the inside of a local system is free, that is, indeterministic, but when it confronts the outside, it is determined to see the curved universe. Thus, LT is made and experienced by the self-centered observers and "located" inside the *minds* as "now time." [18]

## *Rhythm Based Time (*RBT*)*

A Canadian biophysicist, Peter Beamish, has defined Kitada's theory of the local time (*LT*) to mean "rhythm based time" (*RBT*) in biophysics. He has christened axiom 3 of *LT* to KitadaTime (*KT*):

> *KT* = *RBT* as a "local clock": $exp\ \{i\ (+/-2\ \pi)\ Ht/h\}$,

---

is an eigenfunction of the negative Laplacian. See the detailed mathematical procedure and formulations in Kitada, 2002. *Time XI* (http://www.kitada.com/)

[18] It seems that "local systems" of Kitada are quite similar with the windowless "monads" of Leibniz. For Leibniz, as itself unreal time was regarded as a possibility of the temporal location.

where (1) i (imaginary) appears to be simply a real 90 degree *neutrino rotation vector*, within one's mind, (2) radians *+/– 2ð* appears to be simply a *neutrino spin vector* in one's orthogonal mental plane, a plane created by the *i*-rotations, (3) *H* is Hamiltonian transducer which then converts mass/energy neutrino geometry into *KT*: OnTime, LateTime, OffTime, and EarlyTime, followed then by OnTime, once again, (4) *t* is conventional Einsteinian clock-time, and (5*) 1/h* (the reciprocal of the *Planck constant* = 1.5 x 10^33/joule sec) divides a wave function into the *quanta* of mental thought and memory.

Beamish has applied *LT* to his research of cetaceans, and modified it to *cyclic* and *bidirectional "TIME"* = *"NOW TIME"* = *KT* + *ct*, where *KT* is a local clock of one's Event-Space Sphere or Spheroid (*ESSOS*) as a foundation of one's conscious and unconscious mind, and ct is conventional time. *KT* and *ct* are *scalar quantities* and as such cannot be combined. However, if one assumes that *KT* and *ct* are also *scalar tags* that attach either mentally or physically to mass or energy vectors, then *"vector ct"* and *"vector KT"* can be defined with their temporal magnitude and "spacial direction" of quantum mechanical mass and/or energy, and they can be combined by the simplest form of vector addition. The scalar tags are "links" between a local clock, where they are created, and a mind, where they are absorbed, interpreted and used. Thus, *KT* + *ct* as *mental vector protocols* (MVP's) are "NOW TIME" of a mind.

According to Ceta-Research, bidirectional and cyclical *RBT* is the encoding mechanism for "rhythm based communication" (*RBC*), which may allow a possibility to partially merge biophysics and emotional intelligence (*EI*), behavior and health, biosciences and altruism as well as other human endeavors in the new millennium. There are three kinds of communication in nature based on time: (1) communications encoded in *ct* with signals, signs, and symbols, (2) communications encoded in both *ct* and *KT*, and (3) communications encoded in *KT* = *RBT*. Humans appear to live in (1) and (2), that is, mainly with "facts" and "feelings," whereas some low stress animals, for instance cetaceans, appear to live principally in (2) and (3). Switching from *ct* to *KT* requires and produces "low stress", which is an important factor in protecting and enhancing altruistic life in our biosphere.[19]

[19] See BEAMISH 1993, 2002. See also his other papers at: http://www.animalcontact.com/research/research.html#ps.

Cyclic and bidirectional RBT can be illustrated as following declarative and interrogative event-space symmetries:

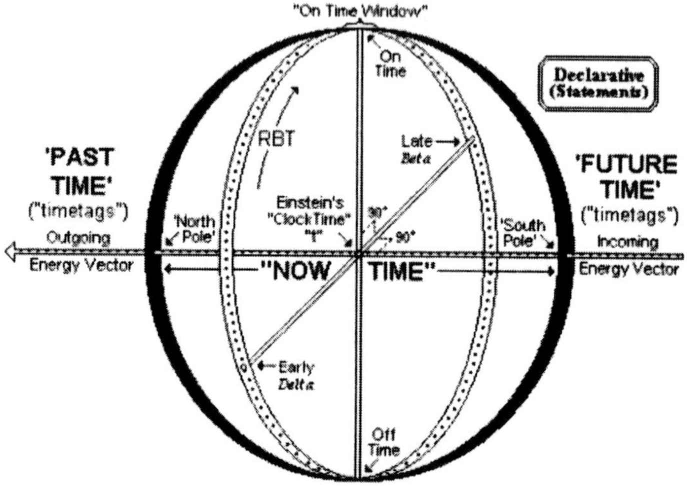


Event-Space Symmetry and "Rhythm Based Time" (RBT)

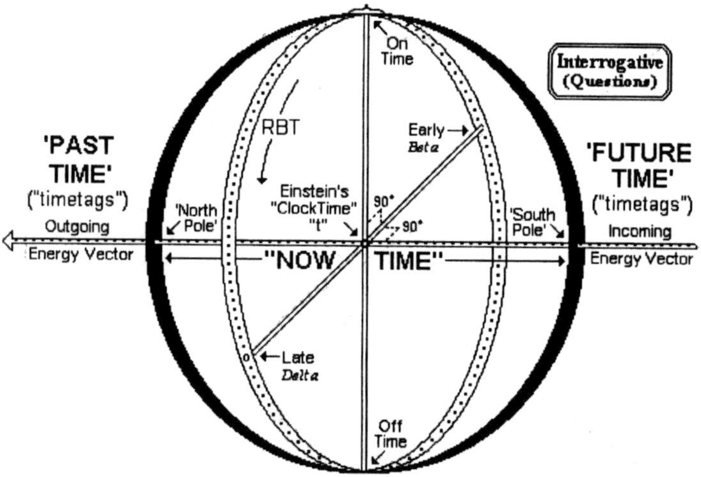


Event-Space Symmetry and "Rhythm Based Time" (RBT)

There are ten variables shown on the diagram as: (A) the *equatorial windows* (1): On Time, (2): Late, (3): Off Time, and (4): Early, (B) the *polar windows* (5): North, and (6): South, whereby energy enters into an organism's Event-Space Sphere or Spheroid (*essos*), (C) the three and only *three orthogonal spacial dimensions* (axes) as variables (7), (8) and (9), and finally, (D) variable (10): classical, scalar, *clock time t*, which always rotates into itself and not into space.

"NOW TIME" is related to the diameter of such an essos. Future time, lying outside of one's essos, is comprised of invented mental labels or *timetags*. Similarly, past time is comprised of mental, inscribed or machine made timetags. However, these scalar labels, real (some of past time) or invented (all of future time), *are not* "TIME," where the latter is composed of mental vector protocols (MVP's) on and within one's essos. Thus, it is derived the important concept: *ALL TIME IS "NOW TIME,"* where "TIME" = "time" (*ct*) + "Time" (*KT*). One can "measure" *ct* as a "timetag difference" (duration) *on* a clock at *essos* edge and mentally "read" *KT* as a "Timetag difference" *of* a clock at essos centre.[20]

## *Physical and eternal time of Heim's dimensional theory (*HDT*)*

A German theologian Karl Heim (1874-1958) defined time ontologically, without any mathematics, although he worked with Georg Cantor in Halle, because theories based on mathematics are either inconsistent or incomplete, as Gödel had proved. Heim's dimensional theory was based mainly on Kant's

[20] See http://www.animalcontact.com/research/ps/ps112.htm. –*RBT-RBC* theory seems to be a novel biophysical variation of time in the existentialism of Martin Heidegger (1889-1976). A basic idea of both of these is similar, namely that time is *bidirectional*. According to Heidegger, there are two types of time: ontic (inauthentic) and ontological (authentic). (1) Ontic time is *conventional* clock-time. It is something to which a "now-point" may be arbitrarily fixed with respect to two other time-points (one earlier and one later), and yet no now-point of time is privileged over any other. (2) Ontological time is "personal": "I am now," that is, my existence is time. "I am" (*Dasein*) means "being-in-the-world," and it is at the same time "being-with/for-one-another" by dialogue or *communication*. My *Dasein* is running to its "past," which takes everything with it into the "nothing." The past is not a "what" but a "how of my *Dasein*," and it becomes visible in "its presentness." In its being the future, time comes back to its past and present. Like cyclical and bidirectional time of *RBT*, Heidegger's ontic time runs forward, but ontological time runs backward. *Dasein* "is" time, not "in" time. Maintaining myself alongside my past in running ahead, I "have" time. Being futuristic as a possibility of *Dasein* I "give" time, because I "am" time itself. "What is time?" became the question: "Who is time?" See HEIDEGGER, 1992.

transcendental idealism, Schelling's (1775-1854) ideal-realism, Heidegger's existentialism, Einstein's SR and GR, and QM. [21] As such, it (*HDT*) is simply a *realistic* version of Kant's transcendental idealism.

The basic idea behind *HDT* and *LT* is similar. There are *two variables of time* inseparable from each other: physical and eternal or timeless. Based on this idea, from the view-point of metaphysics, it is possible to combine two opposite stances toward reality, namely phenomenalism (idealism) and physicalism (realism). From the view-point of physics, the idea allows unification two incompatible conception of time, namely Newtonian absolute (idealistic) and Einsteinian relative (realistic) time. Thus, *HDT* as a combination of two variables of time unifies ontologically GR and QM.

In entire reality, there are limitless or infinite *objectifiable* spaces (*Räume*), which can be defined as well-founded sets. The objectifiable *consciousness-spaces* of my-, your-, and their combination, our-space construct the objectifiable, physical aspect of reality: G-reality (*Geworden*). In other words, there is an objective reality constructed by "many worlds." [22] This objective G-reality is *relative*, physical time. [23] It is impossible to talk about time in general, but only time of individual observers: "my-time" of my-space, "your-time" of your-space or "our-time" of our-space depending on how we are moving through space-time.

Physical time as temporality is the variable of the measurable, countable, and computable change. It is an ever-changing *process*. The future of possibility and potentiality is becoming the present of actuality, and the presence is changing into the past of necessity. The relative flux of time as a "secondary becoming" or a process is an irreversible sequence of successive present moments ($t_1$, $t_2$... $t_n$). Thus, physical time does not start at $t_0$, but at $t_1$, because physical time can be infinitely long or infinitely short but *never zero*, that is, $G > 0$. It means that we can observe only the past of time. The events we observe

[21] See details concerning the philosophical background of Heim's dimensional theory *TDT*, 23-50.

[22] *HDT* is very similar with MWI (many worlds interpretation of QM). See f. ex. Vaidman, http://lanl.arxiv.org/ abs/quant-ph/?9609006.

[23] Heim argued that time is "relative space" in his first publication *Psychologismus oder Antipsychologismus* (Berlin: C.A. Schwetschke und Sohn, 1902). He also argued that time must be "two-dimensional." In my dissertation, *Time and Polarity. The Dimensional Thinking of Karl Heim* (Helsinki: Yliopistopaino, 2000) published as an internet version at: http:// ethesis.helsinki.fi/julkaisut/teo/syste/vk/eerikainen/, and in my book *TDT*, I have reinterpreted and rethought Heim's ideas of his dimensional theory as a novel theory of physical and eternal time.

lie on what is called our past light cone.

Like "clock-time" of SR and GR, "physical time" of *HDT* is relative *space* or at least inseparable from Riemannian metric space and gravity in it. The crucial difference between Einstein's and Heim's thinking was the question concerning the now. The "now" was, for Einstein, the mysterious question beyond physics. He was right, but because he did not want to answer this crucial question, his conception of time was insufficient. According to Heim, this "mysterious something" is W-reality, which as non-objectifiable and eternal time is beyond scientifical inquiries, as Einstein clearly understood.

Thus, there are also limitless or infinite *non-objectifiable* spaces, which are incomputable, and hence, they cannot be defined as any kinds of sets. The non-objectifiable *mind-spaces* of I-, You-, and their combination, We-space construct the eternal aspect of reality: W-reality (*Werden*). It is the *absolute* "now": *nunc aeternum*, that is, $W = 0$. W-reality is the pure non-objectifiable, emerging, and energetic state within G-reality. Heim argued that W-reality does not include only human minds but also non-objectifiable "minds" of the whole biosphere and the material world. For example, the core of electron seems to be such a mental property. In other words, eternal W-reality is *panpsychic*.

Eternal time is the variable of the "primary becoming" or *transition*. It is a certain "imaginary *zero*-point," where the "secondary becoming" or process appears into existence. In other words, "not yet become" *potential* reality becomes "already become," that is, logically necessary actual reality. Entire reality is a dynamic *interaction* and an information exchange of physical and eternal time. Eternal W-reality of the minds or knowing, willing and evaluating subjects (I-, You-, and We-space) is the perspectival and energetic center of physical G-reality (my-, your-, and our-space) of the objects. Eternal time is absolutely simultaneous with all moments of physical time, as St. Thomas Aquinas (1225-1274) thought. Or in other words, at its deeper level, reality is a sort of "super hologram," in which the past, present, and future exist simultaneously, as for instance, David Bohm has proposed. [24]

Even if spaces are limitless, there are boundaries within and between spaces. The *boundary of contents* prevails within a space. If the number of dimensions in a space is n, then the number of dimensions in the boundary of contents is n-1. Instead, the *dimensional boundary* is the constant *velocity of light* = *c* be-

[24] PRATT, "David Bohm and the Implicate Order" in http://www.theosophy-nw.org/theos-nw/science/prat-boh.htm.

tween spaces of W- and G-realities. It means that within a black hole, from where light cannot escape, or within the singularity of Big bang, G-reality of physical time merges into W-reality of eternal time, and hence, $W + G = 0$.

Except the boundaries, there prevails also the *polarity*, that is, the *absolute simultaneity* between W- and G-realities and their spaces. The law of double polarity: *A<>B<>AB* (indifference condition), as Schelling defined it, is neither a causal (cause-effect) nor logical (premise-conclusion) relationship but is obvious immediately. Since Aristotle, philosophers have concluded that if there is no entity that is purely "in itself" and "through itself" (*AB*), then there is no secondary and dependent thing imaginable, and all actuality as a whole dissolves into illusion. The polarity implies that there is no physical time without eternal time, and no eternal time without physical time. *The polarity holds physical and eternal time together.* In other words, reality is not dualistic but monistic totality combined by objectifiable and non-objectifiable aspects of reality. Practically, the polarity means, for instance, that "here" and "there" is absolutely simultaneous in eternal time of I-space but relative in physical time of my-space. In the eternal now of I-space, I can be "here" and "there" simultaneously but not in physical now of my-space.

Thus, eternal time is bound in certain *locations* in physical time, which can be described by using Boolean algebra.[25] Eternal time is like the *Boolean duration* between two measurable "clock-ticks" of clocks. The Boolean non-numerical, *uncountable* duration and the *countable* duration in physical time are always in the *polar* relationship:

$$(\oplus)\ 1<\oplus>2<\oplus>3\ldots<\oplus>n\ (\oplus) \geq 0.$$

Because eternal time exists *within* physical time, *time as a dynamical unity of physical and eternal must be a continuum.* Hence, Heisenberg's uncertainty principle is not valid. Thus, the enigmatic questions of St. Augustine (354-430): "How we can measure time, which is actual only in the present, where there is no duration?" and of McTaggart's: "How can time change in time?" receive their novel answer.

We can conclude that the definition of time would be:

$$\mathbf{G+W \geq 0,\ and\ G \times W = 0,}$$

where G is physical time and W is eternal time.

[25] Reality can be described mathematically, but reality is not constructed or created by mathematics, which is only a human metaphysical intervention toward reality, and as such an intelligent but incomplete approximation of existing reality.

The ontological HDT can be illustrated as a polar *interaction* and an *information exchange*:

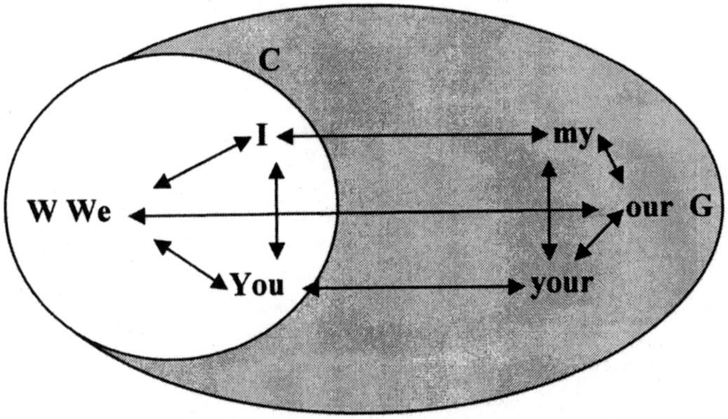


All polar spaces are limitless or infinite. The dimensional boundary = the velocity of light c (circle) prevails between W- and G-realities.

Polarity = ⟵⟶ G-reality = ● W-reality = ○

The *process* of physical time and the *transition* of eternal time can be illustrated as following:

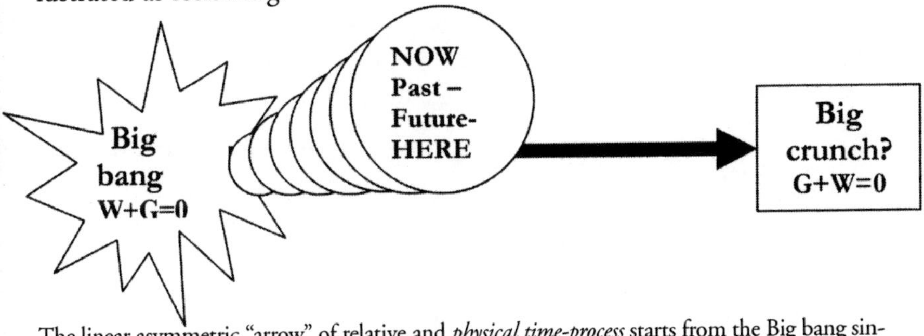


The linear asymmetric "arrow" of relative and *physical time-process* starts from the Big bang singularity (*W+G= 0*), and it may end in the Big crunch singularity (*G+W = 0*). Absolute, vertical, and *eternal time-transition* is present in every moments of linear or horizontal physical time, and so, entire reality is a dynamic interaction of events and an information exchange of physical and eternal time. In physical time, all changes are *successive*, but in eternal time, all happens *at once*.

Thus, time is physical and eternal. Time is in eternity, and eternity is in time.

## *Bibliography*


BEAMISH, PETER

1993 *Dancing With Whales*. New Foundland: Creative Publishers.

2002 *"TIME" beyond "conventional time" II*. Canada: Rodway's.

DAVIES, PAUL

1996 *About Time: Einstein's Unfinished Revolution*. New York: Touchstone Books.

EDDINGTON, A. S.

1929 *The Nature of the Physical World*. Cambridge: Cambridge University Press.

EERIKÄINEN, ATSO A.

2000 *Time and Polarity. The Dimensional Thinking of Karl Heim*. Helsinki: Yliopistopaino.

2003 *Two Dimensions of Time. The Dimensional Thinking of Karl Heim. An Ontological Solution to the Problems of Science, Philosophy and Theology.* Frankfurt a.M.: Peter Lang.

FEYNMAN, RICHARD P.

1997 *Six not-so-easy Pieces. Einstein's Relativity, Symmetry, and Space-Time*. Reading (MA): Perseus Books.

HAWKING, STEPHEN

2003 "The Beginning of Time" at: http://www.hawking.org.uk/lectures/index.html/.

HAWKING, STEPHEN & PENROSE, ROGER

1996 *The Nature of Space and Time*. Princeton (NJ): Princeton University Press.

HEIDEGGER, MARTIN

1992 *The Concept of Time*. Transl. by McNeill, William. Oxford: Blackwell.

HEISING, JAMES W.

2001 *Philosophers of Nothingness. An Essay on the Kyoto School.* Honolulu: University of Hawaii Press.

HEIM, KARL

1902 *Psychologismus oder Antipsychologismus*. Berlin: C. A. Schwetschke und Sohn.

1975 *Die Wandlung im naturwissenschaftlichen Weltbild*. Wuppertal: Aussaat Verlag.

1976 *Der Christliche Gottesglaube und Naturwissenschaft*. Wuppertaal: Aussaat Verlag.

1985 *Glaube und Denken*. Wuppertal: Aussaat Verlag.

JAKI, STANLEY L.

1999 *Means to Message. A Treatise on Truth*. Grand Rapids (MI): William B. Eerdmans Publishing Co.

KITADA, HITOSHI
1999 "A possible solution of the non-existence of time." http://arxiv.org/PS_cache/gr-gc/pdf/9910/
2002 *Locality and the Universe. Time XI.* www.kitada.com.

KITADA, HITOSHI & FLETCHER, LANCELOT R.
1996 "Local Time and the Unification of Physics Part I. Local Time" in *Apeiron*, Vol.3, No.2. 38-45.

LE POIDEVIN, ROBIN & MACBEATH, MURRAY (eds.)
1993 *The Philosophy of Time*. New York: Oxford University Press.

MCTAGGART, J. M. E.
1908 "The Unreality of Time" in Mind, 17: 457-474

PRATT, DAVID
2002 "David Bohm and the Implicate Order" at: http://www.theosophy-nm.org/theosnw/science/prat-bohm.htm.

SCHLIPP, P. A.
1963 *The Philosophy of Rudolf Carnap*. La Salle (IL): Library of Living Philosophers.

VAIDMAN, LEV
2000 "On Schizophrenic Experiences of the Neutron or Why We Should Believe in the Many-Worlds Interpretation of Quantum Theory. http://lanl.arxiv.org/abs/quant-ph/?9609006.

VEZZOLI, GARY
2002 "Physical Consequences of a Momenta-Transfering Particle Theory of Induced Gravity and New Measurements Indicating Variation from Inverse Square Law at Length Scale of .1 mm: Statistical Time Properties of Gravitational Interaction and Analysis Thereof. http://arxiv.org/abs/physics/0104018.

# Autores & editores

Professor Dr. JOHN D. BARROW, Great Britain, Professor of Mathematical Sciences at the University of Cambridge.

Professor Dr. ALFRED DRIESSEN, Netherlands, Professor at the University of Twente, Faculty of Electrical Engineering, Mathematics and Computer Science.

Dr. Th. ATSO A. EERIKÄINEN, Japan, Professor in Philosophy of Religion and Science at the Metasciences Academy, Tokyo. Researcher in the Academy of Finland research program, "Theology and the Concept of Reality."

Professor Dr. KARI ENQVIST, Finland, Professor of Cosmology at the University of Helsinki.

DI. Dr. Tech. CHRISTIAN FUCHS, Austria, Member of the research project "Human Strategies in Complexity: Philosophical Foundations for a Theory of Evolutionary Systems" funded by INTAS (The International Association for the Promotion of Cooperation with Scientists from the New Independent States of the former Soviet Union) and the Austrian Federal Ministry of Education, Science and Culture, lecturer at the Vienna University of Technology.

Dr. Phil. INGA R. GAMMEL, Denmark, Visiting senior researcher in the Academy of Finland research programme, "Theology and the Concept of Reality". At present, visiting senior fellow at Institute of Advanced Study, University of Bologna.

Professor Dr. WOLFGANG HOFKIRCHNER, Austria, Professor for Internet and Society at the University of Salzburg. Associate professor of Technology Assessment at Vienna University of Technology.

Dr. Phil. TARJA KALLIO-TAMMINEN, Finland, Researcher in the Academy of Finland research programme, "Theology and the Concept of Reality." Visiting lecturer in theoretical philosophy at the university of Skövde.

DI. BERT KLAUNINGER, Austria, Member of the research project "Human Strategies in Complexity: Philosophical Foundations for a Theory of Evolutionary Systems" funded by INTAS (The International Association for the Promotion of Co-operation with Scientists from the New Independent States of the Former Soviet Union) and the Austrian Federal Ministry of Education, Science and Culture. Post-graduate student of computer science at the Vienna University of Technology.

Lic. Phil. AHTI LAMPINEN, Finland, Researcher in the Academy of Finland research programme, "Theology and the Concept of Reality." Joensuu.

Lic. Theol. LASSI LARJO, Finland, Post-graduate student at the University of Helsinki. Researcher in the Academy of Finland research programme, "Theology and the Concept of Reality."

Professor Dr. RAIMO LEHTI, Finland, Professor em. of Mathematics at the Helsinki University of Technology.

Professor Dr. EEVA MARTIKAINEN, Finland, Professor of systematic theology at the University of Joensuu. Department of Western Theology. Director of the Research Project "Theology and the Concept of Reality" in the Academy of Finland.

Dr. Tech. TUOMO SUNTOLA, Finland, Adjunct professor at the Tampere University of Technology. Scientific Advisor Fortum Corporation.

Professor Dr. RAINER E. ZIMMERMANN, Germany, Professor of philosophy in Munich and Kassel.